ARTIST'S NOTE:

The figure in the lithograph was inspired by a time in my early twenties and a person who became a mentor to me. Within the image, there is a point alluded to around which his life revolves that is outside normal frames of reference. This makes him isolated and burdened. The horses represent some kind of communication, the way a dream might. In any case, they are wild and overpowering. He's on the edge of something mysterious and intelligent. It appears dark, because it's unknown.
—James Heidt

METAMORPHOSIS:
Transformation In Action

Contributions from the men of

THE EAST BAY NATION OF MEN
THE SAN FRANCISCO MEN'S TRIBE
THE SANTA CRUZ MEN'S COMMUNITY
THE SOUTH BAY NATION OF MEN

METAMORPHOSIS
PRESS

FRONT COVER
ORIGINAL SCRATCHBOARD BY
CLAYTON THIEL
Brujos ♂ *EBNoM*

BACK COVER
ORIGINAL PAPIER MACHÉ SCULPTURE BY
PETER ASHLOCK
Wailers ♂ *EBNoM*

SPECIAL INSERT
ORIGINAL LITHOGRAPH BY
JAMES HEIDT
Bucks ♂ *SFT*

COVER DESIGN
BILL FRIEND
AaaRrrrr! ♂ *SBNoM*

CONTENTS

METAMORPHOSIS

Acknowledgments

It is difficult to imagine any undertaking being supported more extensively or enthusiastically than the creation of this book. From the very outset, people have given generously of their ideas, of their time, of their money, of themselves. In the nine months since the idea for the book was conceived over a casual (way too casual!) conversation, literally hundreds of people have been involved. It would be a Herculean task to acknowledge them all, but in the same spirit that has permeated every aspect of this project, we shall attempt to do so here.

The first challenge we faced was inspiring and collecting the contributions from the 140-odd men whose work is represented in these pages. We heartily thank Tom Blount, Bill Heller, Dave Nehrkorn, Pete Mokris and Moe Rubenzahl for the many hours they spent visiting and cajoling the mostly forthcoming authors, and transcribing and refining the many pages, faxes, dinner napkins and other assorted media in which the material arrived.

Like the Sorcerer's Apprentice, we were overwhelmed by the tidal wave of material submitted, both in magnitude and quality, and realized that a considerable amount of money would be needed to bring the book to fruition. Equal to the task were the community fund-raising teams, who not just met but exceeded the imposing targets they were given. Our hats are off to the South Bay team of "Meta-Heroes" led by captain Mike Druke: P.J. Cotton, Rand Fanshier, Bill Heller, Bill Horst, Jim Knapp, Eric Lane, John Lasersohn, Chip May, Lars Rider, Matt Sorgenfrei and Dave Whitmore; East Bay fund-raising captain Charles Mance with the help of his able-bodied point man Will Scott; San Francisco captain Jay Bunker and the ever-ready Timothy Buchanan; and Santa Cruz captain Gordon Clark and his merry band of Steve Backinoff and Mark Dybdal.

The editing job would probably have been completed some time in 1995 without the help of some consummate wordsmiths. We extend our gratitude to Anne Friend for her skillful review of the raw manuscript, and to Jordan Gruber for his magnificently thorough nitpicking of the galley proofs. We also thank Dave Nehrkorn and Sherry Hagen for their welcome assistance scanning some of the many photographs and art pieces that adorn the book.

The Special Collector's Edition wouldn't be very special without the inclusion of James Heidt's fine art lithograph. Rick Dula of Magnolia Press, Oakland's premier lithographic printer, conceived this enhancement and selflessly carried it through to completion. We salute him and Magnolia's owner, Donald Farnsworth.

We managed to keep within the bounds of legality, if not propriety, due largely to the legal insights of Sheldon Krantz, whom we thank profusely for his time and counsel. Of course we must also acknowledge the folks who actually got the ink onto the paper: Chuck Smith, Roger Heyl and Dick Norman of DD Associates, Inc., our printer, who nursed us through the many design changes with patience and good humor.

Mere words can barely express our gratitude to the many sponsors honored on the following page. For no reason other than their belief in us and the importance of this project, they dug deep and contributed thousands of dollars, without which we would not have been able to get this book into production.

And finally, we offer our most profound and humble thanks to the real heroes of *Metamorphosis*: the men who shared deeply of themselves, and in so doing touched us all forever.

GORILLA SPONSORS—$500

SPARTANS: *Chris Allen, Glen Barlas, Jeff Godfrey, Don Healy,*
Joe Kurlinski, Nolan Mecham, John Pearce

TEAM VICTORY: *Tom Blount, Roger Chapman, Greg Davis, Rand Fanshier,*
Vern McGeorge, Pete Mokris, Ron Tugender

EAST BAY NATION OF MEN

SOUTH BAY NATION OF MEN

RON TUGENDER

RONN LETTERMAN

DD ASSOCIATES, INC.

and
we'd like to thank the BUSHWACKERS

ORANGUTAN SPONSORS—$250

AAARRRRR!: *John Renish, Kevin Hursh, Michael Adler, John Butler, Craig Congdon,*
Ron DeMattio, Michael Dowell, Bill Friend, Sam Zuckswert

QUIET RIOT: *Rick Holden, Donald James, Frank Johnson, Eric Lane,*
John Peterson, Matt Sorgenfrei, Mike Stewart

BOB GRANT

CHIMPANZEE SPONSORS—$100

JIM WILLIAMS	SAN FRANCISCO MEN'S TRIBE	TEAM NFL
KENT SMITH	TEAM PHOENIX	TEAM TATONKA
COLIN GORDON	WAILERS	JUGGERNAUT
DB3	THE NADS — 8/87 — "TI, NS, FU"	

MIKE, SHERI, LAUREL & HEATHER DRUKE

BILL BURTON, *"Honoring the Man who Dug the Well"*

BABOON SPONSORS—$50

BARRY DOBYNS

MICHAEL HOLLANDER

KEVIN LAURENCE KELSEN

DON EDWARDS

JOHN LASERSOHN

DAVE NEHRKORN

WESLEY JOHNSTON

RICK HOLDEN & SANDY MOLL

ANDREW CREELY

KENN D. KEITT

LYNNE PERRY MCCLAIN — OCEAN ARTS, INC.

STARFLIGHT SPACE TECHNOLOGIES, INC.

JERKS OUT OF HAND

GRUNT 'N' POKERS

SPIDER MONKEY SPONSORS—$25

PETER WARTHOG ALEXANDER

TIM WARTHOG BATESON

BOB WARTHOG BLOOM

RICK WARTHOG DULA

DUNCAN WARTHOG GORDON

PETER WARTHOG GRADJANSKY

HOW SHAW

RICHARD KANANEN, *father of Nathan & Steven*

NINA YOUKELSON

SAM MCCLELLAN

DONA TURNER

STARGAZERS

VOHOOSH—APRIL '86

Contributing Communities

Four men's communities have contributed to the content and production of *Metamorphosis*: the East Bay Nation of Men (EBNoM); the San Francisco Men's Tribe (SFT); the Santa Cruz Men's Community (SCMC); and the South Bay Nation of Men (SBNoM). These communities provide a support network for men to experience the unique bond that is part of every man's being. Much of that support occurs within teams, typically groups of four or more men who meet regularly. All contributors to this book are identified with their team and community affiliations. The following four paragraphs, written by leaders in each community, offer a glimpse into the purpose and personality of each.

The East Bay Nation of Men was created in August, 1991, by six men sitting around a table at a Denny's restaurant; the Charter was written on a napkin. Our mission is to support men, teams, and community. The East Bay Nation of Men has grown to a membership of nearly 50 men, most on teams of four to eight men, all supporting each other with passionate and steadfast commitment. Community meetings are held once a month, with an emphasis on outrageous fun, ritual, manly food, individual and team support, and the planning of men's and family events and community service projects. Membership is open to men of any race, creed, or sexual orientation; the only requirement is the desire to have men in your life.

The San Francisco Men's Tribe is a community of men in the Bay Area dedicated to being together in supportive ways. We meet regularly (generally weekly) in teams of four to ten men, and we get together in larger tribal events every six weeks. We also conduct yearly campout events which include initiatory experiences. To participate deeply in the lives of other men is an experience of community that seems absolutely necessary in these times. Every man who allows himself this experience has the opportunity to have a great time while finding out more fully who he is. This is a source of great strength. This community experience starts in the teams and radiates out through the Tribe into the larger world. We welcome any man in the Bay Area who is willing to participate.

The Santa Cruz Men's Community was formed in November, 1981, by one four-man team. At the time, it was an adjunct of the Monterey Community; within a year, enough teams had formed in Santa Cruz County to split off from the Monterey group. In the intervening dozen years, the group has spent thousands of hours in community service, including building a community center and a library; working on half a dozen public schools; cleaning up low-income housing; and participating in a massive earthquake relief effort in 1989. Fun was not neglected, however; we organized numerous river trips, camp-outs, formal dinners, and dances. Today the SCMC still flourishes, using the same founding credo of 1981: To live our lives as men at our best.

In the South Bay Nation of Men, our main focus is leadership development. We realize that the old models of leadership are outdated and simply do not function in today's society. As members of the business world and of our communities, we are committed to developing a new model of leadership and supporting men — all men — to become leaders and examples to other men. As we move toward the 21st century our work is focused on reaching out to the larger community of men to engender a new sense of self-worth and pride. We believe that many of the problems we face in this country arise from a pervasive lack of connection between people. Our mission is to connect men with other men in the age-old bond that will bring men together and say, "You are not alone."

METAMORPHOSIS

Introduction

It is often said that change is the most difficult part of life. A radical shift in one's perception, behavior, or environment takes a person out of his comfort zone and throws him into the great abyss of the unknown. While conscious change — transformation — may be the most difficult part of life, it is often the most exciting and rewarding as well.

Metamorphosis: Transformation in Action represents the deepest, most personal expression of transformation among men in the San Francisco Bay Area. For some men, one single dramatic moment was the transformational event. For others, change was a gradual process. In any case, their work here brings to light the best that men have to offer to themselves, to other men, and to the community at large

The editors of this volume initially put out a call for creative expression "from the gut, from the heart, from the brain, and from the spirit of the warrior." From the quality and quantity of the response, we discovered that we had tapped into a rich and hidden vein: the unexpressed voice of men. This book is the expression of that voice.

Four distinct men's groups took part in the creation of *Metamorphosis*: the East Bay Nation of Men, the San Francisco Men's Tribe, the Santa Cruz Men's Community, and the South Bay Nation of Men. Each of these groups is committed to building a sense of community and cooperation among men.

And cooperation was needed to make this volume possible. Men from these communities donated generously of their time and energy to collect and edit the material; to design the book; to raise funds necessary for printing; to manage the legal, bookkeeping, and technical aspects of the project; and to distribute the book far beyond our communities.

This book is by men, but it is meant for all to read and enjoy.

—The Metamorphosis Team:
Ron Tugender, John Timothy, Bill Friend,
Stuart Krantz & Ronn Letterman
November, 1993

METAMORPHOSIS

CROSSROADS

METAMORPHOSIS

METAMORPHOSIS

Standing in the Window

STEVE GIBSON

Tatonka ♂ SBNoM

The little boy's memory of that May day is very precise. The sky was a beautiful pastel blue with large white clouds slowly dancing across the sky while the warm sun beamed golden shafts of light all around. While standing in the window of his bedroom, his view of the German countryside was pastoral, perhaps worthy of a picture postcard.

His was a typical military family, serving overseas quietly while living in base housing. The apartment building was somewhat isolated from the local populace but the lad traveled off base often enough to savor the feel and essence of the culture. Alien sights and smells assailed the senses.

The little boy was distracted by the beauty of the day. He looked downward and judged the distance from the ledge to the subbasement where access was granted to the rear of the large five story apartment complex. It was with regret that the boy realized that the second floor was not high enough to accomplish his deed, for even though the subbasement added another ten feet, a fall at this level would only result in a broken leg, not death.

He has just been beaten again. Why father used a military web belt and why he was so angry, he has no memory after all these years. He remembers the shame of being in his room; being beaten during a family friend's visit. He remembers that the distance between his room and their location is very short. He does NOT cry. He endures the blows, which are frequent, painful and shameful.

That eleven-year-old boy chose life that day as he turned away from the window and laid face down on his bed. There would be many more beatings and, just as painful, humiliations verbally inflicted upon the youth.

This incident became a metaphor for my life. I chose life that day because I did not live on the third floor. I often think of the differences my death would have made that day: of all the people I would not have met, the children I would not

have sired, of all the men I would not have helped and ultimately, the loss of the love of the woman who has now made life very precious to me. I now choose life every day because it is sweet. I no longer stand in the window.

ORIGINAL PHOTOGRAPH BY
MOE RUBENZAHL
Phoenix ♂ SBNoM

Restlessness — My Enemy, My Friend

RICH SYMMONS
Bushwackers ♂ EBNoM

I am restless today. I am restless most days. And many nights. Sometimes I wonder if my restlessness will ever go away for any amount of time.

I lean on restlessness to get stuff done. I can count on it to be the engine to get me through an all-nighter, or for that matter to power me through many nights with little sleep. God, how I depend on restlessness!

Nothing seems to really make restlessness go away. Sex may make it subside for a while, but sometimes it provokes it. Being involved with something does — sometimes by a silly computer game (although it *must* be challenging!). But

mostly I am relentlessly restless. God, how I hate restlessness!

On one restless night last week, I went in to check on my two sleeping children, two and four years of age. I watched for many minutes as my babies breathed gently in and out. Their ever-so-gentle look made me realize how perfect they are, and what a gift they bring to my life. They became my meditation. My restlessness gave me that opportunity to more deeply appreciate my life.

I am restless so much that I seem to me to be ever vigilant. Why is this so? Why is it so hard for my mind to be quiet? Am I too restless to find out?

Have You Ever?

MARK DYBDAL
Cool Jerks ♂ SCMC

Always his mind on the future
Never where he is or what he is doing

Have you ever noticed how you can have twenty-twenty hindsight but not have a clue during an event as it happens?

Have you ever watched someone you love and known that that moment was going to stay with you as long as you lived, in a photo in your heart and mind?

And have you ever been through a time in your life during which you felt sure your emotions would overwhelm you?

These three questions, as well as any, help me express my personal Metamorphosis from a man to a human.

Ch-Ch-Ch-Changes

BOB BLOOM
Warthogs ♂ *EBNoM*

When was a major shift in my life? Hmmm ... that's something I've never sat and thought about.

What comes to mind? Oh, the usual stuff ... moving to California, changing jobs, that sort of thing. But, are those really *major* changes?

Had you asked me that a year or two ago, those are the kind of answers I would have given ... wait a minute, did I say *would* have given? Maybe I'm on to something! Does that mean I had a major change in the past couple of years?

Yes! All right! I knew I would come up with something if I just let my mind wander a bit.

Changes in my life are happening right now, as I write this, and right now, too, hopefully, as you read this.

What kinds of changes? Well, let's see, most of them are experiential. A year or two ago, I could not have seen myself writing anything creative or personal, maybe not even a month or two ago. Yet, here I am writing.

What started this big change in me? Well, yes, moving to California was part of it. But not just because I moved to California, but because of the events since I got here that led me to take the Men's Weekend and join a men's team.

And what a great team it is! They encourage me, stretch me, push me, piss me off and support me. Sometimes all at the same time!

Wow, I am actually coming up with things to write about! This isn't so bad after all. Is my hand even getting a little tired? Sometimes I can't even write the ideas down fast enough!

I have never really explored the creativity in me. In fact, I used to think that it didn't exist. I then came to realize that we are all creative, and now I am finding, using, and enjoying my creativity.

How?

Well, in the past couple of years I have gone to:

- The Painting Experience, a weekend workshop
- a drum-making workshop
- an introductory improvisation workshop

and I am currently involved in a weekly drumming circle. I've also started to work out on a regular basis. Whitewater rafting? ... never done that, but I will this June.

And there's more that I want to do ... a singing chorus line? a weekly improvisational workshop? Me????

These are some of the things I want to explore more. And I am sure that there is more out there for me to find, do and enjoy.

Gosh, I'm having fun in my life! I don't want to stop changing/stop exploring things.

My suggestion (glad you asked): if you're not exploring all aspects of your life and having fun, don't wait any longer. There's no time like the present!

One Man's View at 44

PETER DAVIS
Guest ♂ EBNoM

ORIGINAL PHOTOGRAPH BY
MOE RUBENZAHL
Phoenix ♂ SBNoM

I used to be young….
I traveled the world
& wrote poems.
I danced the dance of youth.

I used to be young….
I played piano
& sang songs.
I serenaded the lives within me.

I used to be young….
I dreamed many dreams
& fumbled through many loves.
I hoped to become a prince.

I am older now….
I think of money & investments
& developing my children's lives.
I have learned to live with pain.

I am older now….
I still take a turn on the dance floor
& swim the pool length under water.
But mostly, I just sit and watch.

I am older now….
I still hope to get happily lost on
some South Pacific Island, with beer
& friends & native lovers.

I am older now….
I still have some dreams,
But the King is calling me
in the distance.

Breaking Away

STUART KRANTZ
Golden Gate Jerks ♂ *SFT*

All hell broke loose last year. Well, I don't know if I would call it "hell," but something enormous and with intense power filled my body and forever changed the course of my life.

It all started when I began talking about my father in team meetings. I know it has become somewhat cliché to talk about fathers, but I assure you my life will never be the same since these events. Prior to these discussions, my life could be described as a series of events — exciting ones, indeed. An outsider looking in on my life would have been amazed at the relative fullness experienced by such a young man. Travel, education, and excitement.

Until last year, however, I hadn't realized how unfulfilling my life had really been. First I started to feel guilty. Am I allowed to question my privileged life? You bet!

The best way to explain what happened within my body is to make a comparison. Consider two people going through similar experiences. One person, however, is living these experiences through a lens of guilt, duty, and fear. The other person examines guilt, duty, and fear, but does not let these feelings determine behavior. Instead he feels empowered to create his own "lens," one in which allows for maturity and growth, responsibility to community and family, and an overall sense of contentment and well-being.

It was clear right from the beginning that if I wanted to make a bold move in my life, I needed to resolve my anger with my father. But how? Fear set in. Fear of confronting the man I didn't know. Fear of losing the attention I never really received. Fear of appearing weak. But I had my team now, and "family" took on a new extended meaning.

I confronted my dad in El Paso, Texas. Yes, "confront" is the right word. I sat him down and told him to just listen. And I talked for what seemed like two straight hours. I spoke with a passion I never thought I had. Emotions burst out in a stream of consciousness dammed up for over twenty years. He listened. I had his attention.

I forgave my dad that day. I also broke away from him. I am now able to love him without expectations. But even more importantly, I am in touch with a passion that has enveloped my entire being. That day in El Paso began a life-long journey of passion and personal growth.

For My Queen

MATT SORGENFREI
Quiet Riot ♂ *SBNoM*

We are wrapped together
    like the shells and the sand on a beach

Our lives will change us
    just as we will change our lives

And you shall name and rule
        a kingdom within me
    just as I within you shall reign

As the moon rises above the fog
    to lay its glitter on the surf

So shall our love raise our passion
    to trace the map within the stars

And we shall follow that path laid before us
    before we were born

Wake-Up Call

NOLAN MECHAM
Spartans ♂ EBNoM

"*RRRinggg!!!* Hello; American Contractor.

Mr. Mecham?

Yes.

This is Mrs. Scudder at East Avenue Middle School.

Oh, hi, Mrs. Scudder. How are you?

I'm fine, Mr. Mecham. Do you smoke?

Yes, I do.

Do you smoke Winston 100's?

Yes, as a matter of fact, I do.

And do you light them with "Fortune" wooden matches?

Yes. Let me guess ...

Mr. Mecham, Jenny was caught with —

She stole my cigarettes!

Well, yes, it would seem ...

Why, that little ...

I just wanted to get the facts before I confront her. She is waiting outside my office right now. Would you like me to call you when I'm done talking with her?

Yes, I would like that very much. I'd like to talk to her as well.

I'll call you back in a few moments.

Thank you, Mrs. Scudder."

⌾

"*RRRinggg!!!* Hello; American Contractor.

Dad?

Hi, Jenny. What's up?

Mrs. Scudder said you wanted to talk to me.

Yes, I did. You stole my cigarettes.

Uh, yeah.

How many times have you stolen my cigarettes?

Uh ... I don't know.

How long have you been smoking?

A couple weeks or so.

You are too young to smoke.

I know.

Smoking will kill you.

It'll kill you, too!"

⌾

Two phone conversations that changed my life. I realized there was no way I could convince my eleven-year-old daughter to stay away from cigarettes as long as I smoked. Her older sisters both smoked, I smoked, Doreen smoked. I had to take a stand.

As I contemplated my dilemma, smoking what was to be my last cigarette, I fumed that she had pushed me to a decision I had put off for so long. I loved to smoke. I was very angry that she had put this issue in my face. I always knew I would quit, but in my own time.

METAMORPHOSIS

I was haunted by the story about the mother bringing her son to Mahatma Gandhi so he could tell the boy not to eat sugar. Gandhi said, "Come back in three days, and I will do as you ask." When they returned, Gandhi told the boy to stop eating sugar. The mother asked the old guru why he didn't just say it when they were there the first time. Was he testing her? "No," he explained, "Three days ago *I* was still eating sugar."

A few weeks later I asked Jenny if she was still smoking. She said no. I then asked if she had noticed that I was not smoking. She said yes, she had noticed. Did she know why I had quit? Yes. I told her that quitting was one of the hardest things I had ever done. I did it so I could tell her not to smoke.

It was three weeks of pure hell and continues to be difficult. I realized that nothing I ever say would mean anything to me or anyone else if I didn't have the *cajones* to do this thing. To quit.

I Am Strong

JOHN D. STAMM
Tatonka ♂ *SBNoM*

The San Jose sun was blazing hot as it poured through the black window frames, illuminating bright orange parallelograms spaced every five to six feet down the length of the long, narrow dojang (karate studio). I sat on my knees in the third row, second from the end, in my dolbolk (karate uniform) clung to me, drenched thoroughly in my own sweat. It had been the kind of workout that wrenched your guts and emptied your mind of anything except concentrating on what you were doing in the class. I had personally put as much effort into this class as possible. After all, it was the red/black belt class and I wanted to look good among my peers.

The second instructor commanded, "Turn to the right; straighten your uniforms." I responded, feeling kind of feeble. "Face forward," the second instructor again commanded. The class reacted as one. "Salute!" Uniforms snapped with a knife-edge winding up over the heart. "Meditation." We all bowed our heads, striving to relax and let go for several minutes.

Quan Ja Nim (the Head Master) walked up with a big frown and asked in a heavily accented, stentorian, "Who is strong?" No one replied. Not even the more advanced students who had often been admired for skill, strength and endurance. With hesitation, I looked around. I noticed the dust motes floating in the bright sunlight. I replied, "I am," but not very loudly. The Head Master turned and faced me and said, "Whoever said that, stand up." He looked at me and chuckled and said, "You are strong." He told all the other students to do more push-ups. We all went to assume the position, but the Head Master stopped me.

I learned an important lesson. I was empowered by this declaration. I have often remembered this moment when I've been working, ball-handling, or in a few rare moments that it's been necessary for me to defend myself or my family. It was a turning point in my life, as I gained self-esteem and increasingly looked for independence, and relied on myself to handle my problems — because I was strong.

From Curing to Healing

JAY BUNKER
Bucks ♂ SFT

When I began preparation for chiropractic school in 1989, I was driven by a combination of a hunger to learn, desire for more professional status, and a need for greater income. What occurred over the next five years was an initiation which took me back to my roots as a structural body worker and brought me full circle (or rather, full spiral) to being a healing facilitator. And, as with all true initiations, the process was deeply challenging — especially in calling me to question who I am and what I am here for.

As it turns out, the joy of learning can only carry you so far in the face of 30 hours a week of dry classroom lectures and literally hundreds of tests. (Unlike medical school, state law requires attendance be monitored in chiropractic school.) To justify the big time indignity of being a student again at age 36, I fantasized about the increase in capability and recognition I would receive as a doctor. I rationalized that I was willing to trade in my relatively good lifestyle as a body worker for the future promise of greater success. I would gain more acceptance and acknowledgment in the community because I would "fit in" better. Where as a body worker my focus had always been toward service, I was now looking more from the point of view "what's in this for me?" Outside my conscious awareness, however, a greater intelligence was at work in me that had other designs.

In my third quarter, I received an inspired letter from Jenny, my oldest friend, about something on the East Coast called network chiropractic. She had been getting these unusual adjustments and was amazed to find her life changing. She felt I would really like this stuff and urged me to check

it out. I had mixed reactions to this information. On the one hand, it seemed like a great possibility — a chiropractic model that empowers people to make changes in their life. But looming larger than this attraction was a revulsion born of my desire to hopefully be understood and accepted by the mainstream. And besides, life-changing chiropractic sounded like some kind of glorified body work. Why would I be spending all this money and going through so much academic torture just to end up where I had started?

After about another year of the daily indignities and stress of student life, I started having intense pain between my shoulder blades with numbness and tingling in my hand. Adjustments from teachers and fellow students seemed to have no effect. I still had mixed feelings about this network stuff but decided to give it a try. After several sessions I noticed a feeling of relaxation but there was little change in the symptoms I was having. I was looking for symptom relief — not a healing process, so I kept shopping around for something that would get me back to "normal." I must have found the cure I was looking for as the symptoms did eventually fade.

Later that year, my good buddy and carpool mate Mark started getting network chiropractic adjustments. Being someone I knew well, it was obvious to me that it was having a major impact on him. Within several months he became more naturally confident, easygoing and humorous. It was as if the "real" Mark that I knew was there was coming out more and more. Yet despite his enthusiasm and my awareness of the profound effect this work was having on him, I was still not ready to get into it myself. I was unwilling to let go of my self-image as the "holistic doctor." I still fancied myself as a

METAMORPHOSIS

sort of health care "jack of all trades." I couldn't imagine that just adjusting people could be enough. And, just maybe, I was afraid of the havoc that healing might bring about in my life.

ß

At last, in March of 1991, the time came for entering the school clinic. The first step in this process was to find a student "reciprocal" so we could practice working on each other before being unleashed on the general public. The obvious choice for me was my best buddy Mark, but another classmate had already made arrangements with him. My second choice, as chance would have it, was Heather — the only other person in our class who used the network model. As I look back on that decision, I thought it was based on her being relatively new in our class and it would be a good way to become friends. (As with most life changing events, something other than my conscious mind was at work here.) We did become friends. And in the process I received fifteen network chiropractic adjustments, including five in one week as we rushed to complete our clinic entrance requirements.

By the end of that week I had a subtle yet clear sense that my life had been altered. I was experiencing greater ease in my body/mind and knew at some level that healing was being unleashed. While being more in touch with ease, I also started becoming more aware of "dis-ease" in my body in the form of joint pain. This led within one month to fundamental changes in my diet that not only eliminated the pain but also brought a greater sense of well-being.

After several months of care the same back pain and hand symptoms I had earlier experienced and not wanted to deal with came back. This time, rather than fight it, I was able to tune into it as information being expressed by my inner wisdom like the joint pain that led to my dietary change. From this point of view, the pain between my

shoulder blades felt like a constriction in my heart. This awareness helped allow me to begin surrendering the armor I had placed around my heart over the years in response to emotional injury and dishonoring of myself. This time, rather than persisting over weeks, the symptoms dissolved in relatively short order.

When the next network seminar finally came around that fall, I was primed and ready to go. Hearing Dr. Donald Epstein talk about chiropractic as a healing practice was music to my ears after years of listening to the primarily mechanistic and reductionistic perspective of school. But what really knocked my socks off was getting adjusted the first night of the seminar in a room with thirty to forty other chiropractors and students at the same time. (A network chiropractic adjustment can last from fifteen to forty-five minutes, which includes time between contacts for you to integrate or process, so one doctor can best work on several people simultaneously.) This was the most dramatic experience of chiropractic I had ever had. I felt waves moving up and down my spine, sometimes coming from top and bottom and overlapping in the middle. My breath expanded incredibly and seemed to both widen and lengthen my body. I felt tears of sadness that turned midstream into crying for joy in awe of the power of life. I watched insights emerge about my life as my mind came into contact with inner wisdom. When that adjustment was over, all I wanted to do was drape over a chair at the side of the room and feel the bliss of being alive resonate through my body. I knew with every fiber of my being that I had come home.

Being adjusted has become a basic tool for me in tapping into the source of life and healing. Although it is rarely so dramatic as that time at my first seminar, it has a very sustaining and restoring effect that keeps me on track with life. It is usually a deeply relaxing and meditative experience in which I tune in to my body and the innate intelligence at the core of my life. I feel patterns of stress in my spinal cord release as the gentle contacts are made on my body, often with subtle wave motions flowing through me. I also love the exquisite releases that can occur with perfectly timed structural adjustments. I have definitely been converted to chiropractic for life.

I realize now that my resistance to network chiropractic was based largely on fear of getting involved in my own healing process. I hoped and assumed that I could help other people heal themselves while maintaining a safe, "professional" distance. I know now that I cannot so easily draw a line between my own process and that of people I work on. Any area I am unwilling to deal with in my own life will block me in working with others. And my own willingness to surrender to the healing power of life is necessary if I am to facilitate that process in others.

When I undertook this journey five years ago, I thought I would end up being more a part of the conventionally accepted "health care delivery team." What has actually occurred is a deeper access to the work I began ten years ago as a body worker. Rather than help people "get rid" of pain and other symptoms, I am working to assist them to trust and free up the healing power within themselves. In retrospect, it seems that the whole process has been guided by my inner wisdom. The goals I set out to accomplish are occurring: greater learning, professional status, and income. But rather than being a result of attaining something outside myself, they are coming about through gaining deeper access to myself. And, it turns out, those justifications I gave myself for undertaking this journey are far outweighed by the inherent joy of discovering the power of our innate intelligence.

There Are No Mistakes

WILL SCOTT
Brujos ♂ *EBNoM*

Twenty-five years ago I was busy climbing as fast as I could up the corporate ladder. As far as I was concerned the accumulation of money and material goods were the prime reasons for being alive. Accomplishments had to be measured physically and materially or they really didn't mean anything. Life was pretty black and white. I was a dyed-in-the-wool materialist and confirmed agnostic who equated spiritualism with religious dogma and refused to have any part of it. My transformation into a believer in the power of love and the human spirit took place without my even being consciously aware that it was happening at the time.

One day in the late spring of 1968 I watched a handful of Buddhist monks caring for nearly 300 orphaned Vietnamese children in an enclave of makeshift buildings outside Long Thanh, almost two hours north of Saigon. In spite of the many hardships, a tremendous lack of adequate supplies and the almost daily influx of more children, those humble monks projected unconditional love toward all as they devotedly went about their business caring for their charges as best they could. The children ranged in age from about a year old to nearly eighteen. The long, low shacks they lived in were built of scrap lumber, with packed-dirt floors and crudely thatched roofs of straw. Whatever possessions the child showed up at the gate with were all he or she would have. There were no mattresses or blankets for them; a plywood slab functioned as a bed and the clothes on their small backs was the only cushioning and warmth the children had to ward off dampness

M E T A M O R P H O S I S

and cold at night. Bathing consisted of dipping each child in an old 55 gallon oil drum filled with water. The monks did their best to raise some crops and the nearby villages supplied some food also, but there was not a lot to go around and food was becoming more of a problem as more children appeared at the gate.

I was in the Ninth Infantry Division, U.S. Army. As a paymaster my job was to spend up to two weeks of every month in the field with the infantry, paying them and handling any problems about allotments, etc. Every month was an ordeal of witnessing more death, mutilation and atrocities and thanking God that I wasn't out there in the infantry full-time. That experience convinced me that the most gentle person is capable of heinous acts against his fellow man if put into the right circumstances. My denial and survival mechanisms prevented me from fully comprehending what I witnessed every month or I probably would have run screaming into the sea. It took the passage of several years before the experience could be digested and regurgitated in a coherent form. Even now it's hard to talk about some of it without becoming emotional.

I am certain now that every experience has something to teach me. The war and the orphanage came into my life to show me that simple acts of love and compassion can prevail even in the midst of hatred and wanton destruction of life and property.

Over a period of three months I had almost weekly contact with the orphanage. It nearly doubled in size during that time, but the monks just continued toiling with the same love and devotion as before. After my first visit there I had written a letter home describing the orphanage. That letter was printed in my hometown newspaper and

subsequently sent to people in various parts of the United States. Within weeks I began receiving packages of clothes, toys, blankets and food from all over the country. Each time I delivered these gifts to the orphanage, the monks would have the children do something for us as a thank you. They would sing in English, French and Vietnamese. Or they would do some traditional dances. The simplicity, beauty and energy of those children still bring tears to my eyes.

For me the orphanage was a breath of fresh air — a place filled with the spirit of children and the love emanating from the Buddhist monks. The war did not exist for me when I was there. It took years for me to truly understand the impact of those beautiful monks on my psyche. The compassion of those monks, and of the Vietnamese people in general, left a deep imprint on my mind. Even though they have faded considerably, the scars of that war still exist within me, but the love I witnessed in that orphanage touched me far deeper. I believe that experience is in large part responsible for me leaving the corporate world and being what I am today — a teacher and healer of the mind, body and spirit.

Without that experience I may never have begun meditating and learning to quiet at least some of the endless monkey-mind nonsense that often fills the space between my ears. And I may never have found my own spiritual path that has led me to a much fuller life than would otherwise be true. I could very well be living a moderately unhappy life in the corporate world, doing my best to convince myself that all I needed was a few more toys and all would be perfect. As strange as it may seem, I thank God for sending me to Vietnam. It may have saved me from a life of mediocrity and spiritual purgatory.

The Mountains

BILL HELLER
More Power! ♂ *SBNoM*

There Came a time, a point of despair in my life.
A time when the most common feeling is Useless.
All I have accomplished is meaningless,
because those things most important to me are not available,
at least the way I think they ought to be.
A time I continue to screw things up.
A time when I am questioning my own sanity.
How can I continue to self-destruct this way?
How can I trust myself?
WHERE DO I GO WITH THIS RAGE!!

⌀

On a rainy August Day in 1990
I am alone on my way to the mountains.
And I am alone, in my pain and suffering.
And I am alone in my fear.

I see no answer.
In my grief,
I scream in tears of rage and frustration,
while driving and attempting to hide under Santana's music full bore.
The feeling is inescapable:
LET IT OUT BILL, IT'S YOUR TIME.

Now in the mountains, all others have left, due to the rain.
I walk along the path, blinded by my tears.
The only sounds are the rain
and my roars, emanating from my gut.

As I continue this rage,
I begin to notice something I hadn't before;
A peace and connection that exists in the woods.
And as the rain continues all around me, none falls on me.
The rain stops.
A dead branch on a living tree nudges me as I walk by.

I AM WALKING BECAUSE I WANT TO, AND NO OTHER REASON.

METAMORPHOSIS

What a novel idea.
A feeling of peace.
A feeling gone for too long.
The Sun comes and goes;
I realize now it has always been there to soothe.

AND I AM NOT ALONE.

A view of the mountains to the west, I am in awe.
I yell Thank you; I realize I can and do love.
The gusting wind yells back to me.
This forest is here for me, and I for it.

Walking again, I kneel and hold my absent daughter, Vanessa's hand.
I grieve for you, and in spirit I am always with you.
Vanessa you are perfect.
Similar thoughts for my children Brooke and Matthew.
I realize they know the connection exists.

I AM COMMITTED TO THE WELL-BEING OF MEN.
FOR THEIR DISCOVERY, IN THEIR TIME.

This moment is perfect.
Here, there is nothing to change.
I AM COMPLETE, AND ACCEPT MY SELF AS I AM.

✍

TRANSFORMATION

The walk started as an opportunity for pain, anger, grief. It led to a feeling of completeness, of connection. A feeling of enthusiasm about sharing with the others in my life. Thus started my return to God. I've since learned the Spirit in the Woods, God, has no ego, no need to judge, condemn or criticize. It accepts me as I am, and loves me completely. I can too, both myself and others.

When leaving the mountains, I noticed a shift. I had rediscovered My Self, and my connection to God. Nothing had been resolved about those "things" I was dealing with. I was given something far more valuable, and with it, a peace, and faith these other matters would be resolved in their time.

This was the start of a rebirth that has continued for almost three years now.

THANK YOU, FATHER.

STAR House Taught Me

Yin-Sun Wu

♂ *SBNoM*

I had no idea, when we started, that it was going to be this hard.

It was a Wednesday evening, in April last year. Our men's team walked into the STAR House group home in Sunnyvale. We were there to visit with the six teenage boys that live in the house — eat pizza, play some cards, have fun. This was our fifth activity with them in as many months. A little over two hours later, we came out shaking our heads, feeling defeated. Two of the boys had disappeared into their rooms right after dinner, one went back to his video game, one back to the television. The two boys that could be persuaded to join us in a game of cards played without enthusiasm and were surly when they didn't win. No one had fun, the evening was a flop. We were disappointed, frustrated, annoyed, and hurt. I said, "Ouch! Tell me again, why are we doing this?"

I knew why. We wanted to establish an ongoing relationship with the boys, to be positive role models, to give them the experience of being around decent, responsible men — the type of men they'd probably seen very few of, if any, in their young, hard lives. (They range in age from 15 to 17, and come from abusive or neglectful homes.)

But this was tougher than any of us had anticipated. We'd been at it for a while now, felt like we'd done a lot. Picnics, volleyball, touch football. A Warriors basketball game. We'd built each boy a new bed as part of a special, multi-team project for all the STAR Houses. I was one of two men who had gone through the certification process to be cleared to drive them on our outings. And yet, each time we got together with the boys their response was the same — a cool greeting, zero eye contact, monosyllabic answers to our attempts at conversation, no feedback whatsoever that what we were doing was having any positive effect. No outright rudeness or rebellion, but no warmth or sense of connection either. Each time felt like we were starting from scratch. I don't think any of us had experienced this kind of indifference and emotional detachment before, not even from bad blind dates or ex-girlfriends ... Hey c'mon, remember us? We're the good guys. How come it's so quiet and cold in here?

We complained to each other. "I'd like to strangle that little turd with his stupid video game," I vented to a teammate. "Yeah, me too. Why doesn't he join in?"

We gave each other positive feedback, and pressed on. "Given their experience of the world, why should they trust us?" "Okay, we said we'd do this for six months, let's do it. Little shits!"

At the end of our first commitment period, our team leader, Pete, said: "This is an open-ended project we've taken on. We're not going to discuss committing to another six months until next week. Let's take this week off, to acknowledge and feel good about what we have done during these first six months." That's a brilliant idea! I thought. Create our own milestones and take the time to celebrate them! It was a great lesson. We all signed up for another term.

The next month, we got the first sign of life from the boys: they requested that volleyball be part of this month's activity — Hmm! We played on the beach north of Santa Cruz — it was a good day.

METAMORPHOSIS

In August, a big setback. At a barbecue picnic, one boy, acting out, punched one of our men from behind, twice. The man confronted the boy pointing out the dishonor in hitting from behind. Later, he spoke again with the boy to make sure he understood that it was his behavior that was objectionable, not him. The boy seemed OK, but there were repercussions. We were told: Do not confront or attempt to discipline any of the boys. Disengage, call the counselor over if an interaction looks like it might escalate into physical or verbal violence. Don't try to teach, counsel or advise the boys about what's appropriate behavior or good sportsmanship. Or about being a man — other than by modeling it. Compliment or affirm a boy only for very specific actions, don't make general statements of approval or affirmation.

Wait a minute, I objected, feeling defensive and ashamed. You mean I wasn't supposed to tell Wayne I see great leadership qualities in him and I think he has great potential?

That's right. His self-esteem and trust of adults is so low that his internal reaction would be "This idiot thinks I'm great. Doesn't he know? ..." Show them they're great, don't tell them.

Aha! I thought. Compliment them for specific deeds and accomplishments. Affirm them by showing up month after month.

But wait. We've done the Men's Weekend. We meet every week to support each other to be men. We think we know a lot about being men. We like to share what we know.

That's right. And your job here is to give the boys an opportunity to experience a healthy and appropriate adult male relationship, through interaction and role modeling.

I got it. Men are defined by their actions, and STAR House is a golden opportunity to test ourselves, the way we show up as men. With these boys, we have to communicate and relate at the most elemental, male level — from the gut and from the heart, by doing, by being. Without the normal social niceties and masks, without the talk, the verbal smoke screens. That's the level the boys are at and that's the discipline for us. I said to Pete, "They do know we're there! They don't look at us, they don't talk with us, but their antennae are up all the time. They get who we are and what we are. We just have to show up and pass the test. Over and over again. What a great practice!"

A tough practice admittedly. Unfamiliar and uncomfortable. The lack of response, lack of any sign of appreciation was disconcerting and disorienting. It hurt every time I went to shake young Eric's hand to have him ignore it and turn away. And I felt sad and hopeless every time I looked at young Bill and wondered whether he would ever be able to move past his enormous anger and hostility towards everyone and everything. But I realized that, even though we don't get to make the rules, if we could stay on purpose, accept these boys as they are and surrender to what works with them, we just might achieve something worthwhile. The truth is, it's highly unlikely we'll ever see the real results of our work. But if one day, on the streets, a boy makes a good decision because of his experience of us, that would make all our effort and discomfort worthwhile.

We continued, and as my teammates and I supported each other through frustrations and uncertainty, I learned I could have patience and perseverance beyond what I thought I'd ever be capable of. That gift will stay with me the rest of my life.

✍

A good step forward in October. Our team (men and boys) won the big basketball tournament, beating the favored STAR House 1 team in the final. A sweet victory — we'd been nervous because our two practice sessions had been busts; the boys were competent individually but seemed incapable of playing as a team. The boys had been *real* nervous, convinced of their opponents' prowess from experience and from the psych-out war of words that preceded the tournament. We won through teamwork. The trophy occupies a place of honor in the boys' living room. They won't let the counselors touch it.

At the end of December, we went bowling with the three boys who had no homes to go to during the holidays. I noticed something that was both surprising and exciting. As each boy bowled, turned around, and walked back towards us, he looked over at us for a fraction of a second. The most fleeting of glances but unmistakable. Now I got ready, and in those split seconds, I tried to be there. If he had bowled a good ball, I gave a firm thumbs up and nodded my congratulations. If it was a bad ball, I gestured "Shucks! Too bad!" Each time, a hint of a smile, a tiny softening of the boy's face told me I had made contact. They're just like me! They don't want to show it (too vulnerable, and besides, not cool) but they want somebody to notice, somebody to take an interest, somebody to share their victories with, somebody to console them when they don't do so well. I, as a man, can do that for them.

On a Wednesday evening last month, members of our team walked into the STAR House group home. The atmosphere is different. The boys are more relaxed around us, a couple of them even cracking jokes. The counselors tell us the boys really appreciate their activities with us. We know we're making a difference. (We've been making a difference all along — it just didn't look like it.) There are three new boys in the house now — they've heard about us from the others and that makes relating easier. We've learned not to have any great expectations, though. After 15 months, we feel like we're getting the hang of it.

✍

The boys' names have been changed.

To the boys and STAR House counselors and staff, many thanks for the opportunity to give and receive. I am humbled by and proud of our association.

My heartfelt thanks go to my men's team teammates. No way I could have learned what I learned without you men. We've been a great team on this journey.

After the Weekend

DAVID HOHL
More Power! ♂ *SBNoM*

After the men's weekend, I had one of the clearest weeks I have ever known. I was totally present. When I heard a song I flowed with it. When I listened to people my inner voice wasn't judgmental. When asked to commit to a job early the next day I said yes without reservation or conditions. I saw someone stuck on the freeway, and I called the police. I felt very much a part of all the people and the world. My inner feeling was one of being a proud lion on the Serengeti plain. Not conceited or egotistical, just very centered and at peace. I will not forget this brief moment in time when the world was new again.

Talk'n Drums—Notes From A Pilgrim

RON DEMATTIO
AaaRrrrr! ☾ *SBNoM*

Who am that bam bam man? According to my mother, I started drumming when I was still in the crib. With her support, I took 8 years of drum lessons starting at age 6. My father always grumbled that it was a waste of money. Even as a kid, I felt my Dad acted more like a jealous sibling (who could hit and abuse) than a father. My father shut me out an early age, and was verbally abusive. One of his favorites was to tell me I was worthless and would never amount to anything. I still hear that voice in my head, usually disguised as my own.

I went to an old-line Catholic school in Pennsylvania that operated on the principle that you were born doomed, and everything you do or think is **wrong** and **WILL** be held against you in a court of God (the vindictive bastard!). When the nuns weren't using my head, hands, or ass as their drum, in an effort to stop me from sinning against God by making duck noises in class, I was home reading science books, drawing, and drumming to records my mom bought for me. No matter how much the nuns or my father tried to squash my spirit, they could never take away the pure, strong and simple joy I got from music.

From all of this, and fights I got in, I "learned" that women had all the power of approval, affection, and judgment, and that men would knock you down, kick you in the head, call you names and laugh about it, and ignore you if you reached out for help. I created my own world, where I had few male friends and constantly sought after women for all the "goodies."

My first public drumming was for Catholic men's retreats, setting the pace for a quiet procession across town, and up a hill to a breakfast and a talk by a bishop and some other holy pancake eaters.

When I was 13, we moved to California. I think I went into warp speed to prove I was not stupid or a loser. I not only was performing in bands earning money at age 13, but was in the advanced band at school, a concert band, marching band, an orchestra and big band jazz band after school. At school, I took as many "solids" (like trig and physics) as were allowed.

I became active in the environmental movement for almost 10 years, where some people challenged me to give up my music and devote all my time to saving the world. After college, in the 70's, I burned out on politics, worked in print shops, got into progressive rock bands, bought a piano and started to express my melodic side. I started acting in local theater, video productions, and training for acting in film/TV with a dream to do voice and music in THE BUSINESS. When the world got heavy, I would pound on the drums for an hour, lurch sweating into my room to play the piano for an hour, and then pass out on the bed. Music gave me such a high, I couldn't understand how people could waste so much time passively on drugs. Whatever was going on in my life, I always felt that *no one could take away my music.*

I also submerged into relationships with women who thought I should "grow out of" being a musician/artist, and become a scientist or engineer. This and a few self-help workshops led me to do the *Men, Sex and Power* weekend in September '85 a few months after my father died. I began to foster communication with ... men, and to feel good about myself in the process.

In '86, I produced and marketed a tape of my improvisational piano music called *Dream Suites.* Doors were opening, seeds were sprouting in music *and* acting, when a car accident caused soft tissue damage in my neck and shoulders, affecting my arms and hands for a long time and giving me intense pain, headaches, and depressions. For most of my life, my greatest fear was of injury to my hands. It seemed that fear had been realized. I stopped drumming, cut back marketing of my music, passed up opportunities to perform, record and act because of my health. My Catholic up-bringing said God was punishing me for being too ambitious. I felt like I had been knocked off the train, a loser. I went headlong into the darkness.

I'm grateful to my team for sticking with me regardless of my level of OK'ness. Actually AaaRrrrr Team is my second team (I joined in '87); my first team died of over-analysis. Being on a team and active in the men's community has kept me sane, and probably alive.

Over a year ago, my team went to a day with Robert Bly, Michael Meade, James Hillman, and about 750 other men. I felt a weight lifting while listening to them talk about the natural and healing process of going through your own darkness, and the power of the masculine in the

arts. Bly encourages men to use the arts to grow and explore your personal spirit/journey — any art: writing, painting, dance, drumming. Many times during the talk I felt like crying; this was like a message from God saying I was still on **my** path.

The darkness I was in was an **important** part of that journey. Inspired by the drumming, my team later bought a pair of congas. They appointed me Keeper of the Drums, and I started drumming again. I've since coached small drum circles, drummed at Nation of Men meetings and events, and have been expanding into other groups. My goal has always been to create a space of magic, passion, and grounding for the men. When I drum, I get a strong feeling of flow, centeredness, and communion, as though I'm communicating spirit from within, nonverbally. Drumming connects me to my inner power center. Even my keyboard music has developed a more upbeat, passionate aliveness. I am painting again, writing poetry, acting, training for voice acting (voice-over), and plan to produce another keyboard album and a drumming album. The drumming and the men in my life remind me to be who I am.

In '92 the SBNoM Chiefs sent me to a drum workshop with drumming master Babatunde Olatunji, shifting my confidence and abilities into higher gear. In the words of Olatunji:

> The language of the drum is deep. It's like your heartbeat. The rhythm touches every cell in your body and mine. It can be used to calm a person down or cool off a situation. It can be used to arouse passion. It can be used to heal — for the simple reason that every cell in our bodies is in constant rhythm.

Up to now, I've played the role of "starving artist." Recently someone said, "Ron, you are a rich man, you just don't have a lot of money!" I heard you!

Listen to your hearts ... beat.

ORIGINAL DRAWING BY
BILL MCCONNELL
Phoenix ♂ *SBNoM*

The Source

DONALD JAMES
Quiet Riot ♂ *SBNoM*

It was 1978. Second semester of my junior year at USC, Spring in L.A. where the nights can be Hawaii warm.

On one of these nights, I was burning the oil working on a paper. Seems to me it was about 2 or 3 a.m. I really wanted to blow off the paper and go to bed. My decision to stay up and finish the paper would have an impact on my life far greater than I could have imagined.

The Shrine apartments were within walking distance from the campus. That's why my roommate and I got the place. Close and cheap. It was located in one of "those" neighborhoods around USC — the ones white folks didn't like because the nonwhite to white ratio was higher than tuition. Typical nights, weren't. Peace and serenity, foreign concepts.

This night was no different. I grew accustomed to the ambient noise: sirens, helicopters, loud music, drunks and that university chime. But through the cacophony of noise this one night, I heard a different sound. At first it didn't strike me as odd. People clapping or cheering — like at a basketball game. But there was no basketball game. Why would there be people cheering at 2:30 in the morning? Strange.

I ignored it at first, but then my curiosity overpowered my commitment to my paper. Just what were people doing cheering and clapping at 2:30 in the morning? I eased onto the balcony to see what I could see. Nothing. The noise seemed to come from down the street towards the Shrine Auditorium. I went downstairs and looked down the street.

What I saw did not lessen my curiosity. People streaming INTO the Shrine — at 2:30 in the morning.

I walked closer, like a cat curious about a new object in his house. There were many people. They were smiling talking loud and full of energy. I watched. I could hear a male voice booming inside the Shrine. Periodically the crowd roared. You could feel the energy in the air. "What the hell is going on?" I thought.

There was a table in front of the door into which these people went. There were three people at this table, seeming to check people back into the Shrine. People kept streaming in. Happy, smiling and energetic — it was now 3:00 o'clock in the morning.

"Excuse me." I said with an of-course-why-wouldn't-I-ask tone. "What's happening here?"

"They're getting it," said the young woman with long black hair.

"Getting what?"

"GETting it."

"I don't understand."

She seemed a bit distracted by my inability to get "getting it." Patiently she tried to explain. I don't recall much of what she said. But I do remember that word she kept using that seemed to be the focus of all this:

"est."

☙

That's it. That was when the whole process, the whole idea, the concept of the birth of transformation began for me. Of course I didn't know it at the time. A few years later a colleague at work invited me to an est guest seminar. I went to about 3 or 4 before I "got" that I was supposed to register to take "the training." By the time I did register it was called "The Forum." That was in April 1984.

The Forum had the greatest impact on my life — ever. I would never be the same. Hearing that word — est — was like the sperm fertilizing the egg. Something was going to be born. And after some years of gestation, something was born — me!

Other life-enhancing events of my life emanated from that experience. I got my wife. I got the Men's Weekend. I got the Nation of Men (or as est founder Werner Erhard might say, "they got me"). My relationship with my family changed. My relationship with work, religion, sex, money and most importantly, my relationship with my "self" changed. The changes have increased the quality of my life by orders of magnitude.

In retrospect I feel even more fortunate. Erhard isn't around leading trainings anymore. But in spite of the problems and issues that have been reported about him, I believe that history will judge him to be one of the most influential and great people of our time. Given what I've "heard" about him I doubt I would have signed up for his "training."

I shudder at the possibility of that. Still in the darkness; water to the fish; air to the bird. And that's just the way life is. No possibility.

I'm glad I stayed up that night and surrendered to my curiosity. Funny how we create life.

There Is No Life After Death

John Butler III
AaaRrrrr! ♂ SBNoM

A man must walk all the way home
There are no half written poems
And no half taken journeys
It must come from all the way within
And it must flow all the way without

As sunrise my journey and my truth came to light
I know what I am called to do

A lady bug sits on a blade of spring wheat
In an hour it will rise and fly to the mountains
As it sits
It does not know how to fly, nor where the mountain is
It only knows it is time to begin
It just knows its soul can wait no longer

A poet is not born of solitude
There is always one whisper
A desperate and quiet source within
That forms a community
That breeds compassion
And begins the flight

Angels

DENNIS SIDES
Happy Men ♂ *SBNoM*

There are many roads to transformation — falling in love, a near-death experience, the birth of a son, getting dumped, etc. I wrote this in October, 1990, after my mother died.

the angels weep for three days and three nights,
as you dream morphine dreams in the cancer ward
you're as solemn as a buddhist monk on a hunger strike,
fevered and yellowish against the white sheets,
a halo of sweat forming around your bare head
earlier today, your tentative hand shakes, barely grasps
a pen to sign the DO NOT RESUSCITATE form you requested
I choke up watching you struggle to make a pathetic squiggle
the rain rolls down my face (I hear angels singing)
as I realize this time you won't be coming home ...

needle jabs have tortured your arms, covering them
with purple bruises, punching the sad little holes
doctors use to drip life back into you,
while it drains from the larger hole they can't find
storm clouds pass across your face, haunted & fearful,
so I say, "Just talk to Jesus, He will hear you —
after all, His mother's name is Mary too," and then
you smile like angels, sad & faraway, beyond conversation ...

at night I stand at the window, search the drizzling darkness
for that glorious golden light I keep telling you is coming ...

The Continuing Transformation of Terrence

TERRY MCMAHON
Bushwackers ♂ EBNoM

"All right, children, line up, boys on the left and girls on the right, that's it, side by side, and stand up straight. Good! Now boys, hold your right hand out as if you were holding a serving tray. Good! Now girls, scoot in close to your partner and lay your hand on top of his. Great! Now let's proceed into the dance hall and learn to dance!"

The year was 1961. I knew nothing of dancing and wasn't real sure that I was going to get used to the idea of holding a girl's hand — YUK! But this was, alas, no ordinary girl: this was Gwendolyn Peterson. Years later I would discover that she had more syllables in her name than anyone in the class. But for now all I knew was that she had a commanding presence and a stronger grip than perhaps I have today.

While we were standing there assuming the position, the dance instructor would go down the line (there were almost 50 students in the class) and make sure everyone was properly coupled-up and positioned. This could sometimes take as long as ten minutes. As the dance teacher did her thing, Gwendolyn would proceed to squeeze my hand to the point where it felt like permanent skeletal and nerve damage was a distinct possibility.

Now this was a very difficult dilemma for a boy of six. My options as I saw it were to:

(a)   pretend this wasn't happening: "No, my hand is really not in a vice; it feels fine."

(b)   reason with her: "Was there something in your youth that made you hostile towards men?"

(c)   tell the dance teacher — yeah, right: "Terry is a Ninny; he got hurt by a girl!" OR

(d)   bribe her: "Listen, when you grow up to be the WWF wrestling champ in high school, I'll still dance with you if you would just stop destroying my hand now."

None of these options were sitting well with me, but I continually chose option (a).

My outlook on life was diminishing, and apparently it was showing because my mom asked me one day after school what was the matter. I told her about my predicament, and she asked me what I could do besides what I was currently doing. I told her that I would like to cause Gwendolyn physical pain in return for the pain she had inflicted on me. My mom asked why I hadn't. I told her that I understood it to be pretty near a mortal sin to hit a girl (I apparently had a tendency to hit my younger sisters before my mom put the fear of GOD [her backhand] in me).

"Well, you can get your point across without hitting her," my mom said. "Stomp on her foot!"

"Seriously, I could do that?"

"Yes," my mom said, "If she refuses to stop, you could stomp on her foot."

My life changed at that moment. I recognized that I no longer had to be a helpless victim of the wrath of Gwendolyn.

The very next time we lined up for dancing practice, we assumed the position, and the vice grip started.

"Please stop that, Gwendolyn," I calmly asked. She grinned and only squeezed harder. I glanced over her shoulder to divert her attention. As she turned her head I raised my foot and slammed the heel of my shoe onto the toes of her shoe.

A brief shriek was followed by bulging eyes and a welcome limpness in her hand. She looked at me as if to say, "How could you?" I just smiled.

That day we danced the best we ever had, albeit with a slight limp on her part.

But from then on she seemed to gain a certain respect for me, and I recognized that I have more options than I sometimes realize.

METAMORPHOSIS

INSIGHTS

METAMORPHOSIS

Nothing Special

MIKE PALMER
♂ EBNoM

Since I climb mountains, it seems to many non-climbers that I would have had a few brushes with the Grim Reaper. This is not as common as many would believe. A major element of climbing is pushing your limits without getting the ax. In other words, knowing what you are doing. This is one occasion where I could have known more.

The road was nearby but this was still a quiet place. Only the light wind through the morning chill of eastside pines made the music that can soothe tense young climbers and give them the feeling of life and strength. Alan and I loped out into this air, down the boulder field to the base of a frozen waterfall. I was charged, ready, pumped for a fight. Alan was unusually quiet. This was our first real ice climb.

Brittle ice fragments tinkled down as we started up. We were too young and dumb to be scared — yet. The first belay was set after only 50 feet of strenuous climbing; 150 feet is more common. We were both too serious to talk much.

The sun was out now. As I led up the next section I could hear water running under the ice. I set my axes as if I were trying to punch through concrete. As the sun hit the upper part of the fall, ice blocks began melting loose from above and whirling by me. I could see water flowing over the rock under the ice as my axes broke small portholes through the veneer. The ice we were on was not attached to the rock beneath it. I was about 150 feet above the deck. No question now, I was scared.

I found some thick ice and placed a screw to protect myself in case of a fall. It was time to get off this mono-lith. It was time to go home. Alan anxiously agreed. We were in over our abilities and beyond our common sense.

I felt like a failure as I moved right from ice bulge to ice bulge, moving towards a ledge on the rock wall where I could set a rappel. Was I a failure? I had not finished the climb. Was this good sense, or was I just too scared too put my ass on the line?

These thoughts raced down the sides of my brain through my shoulders and set like weights in my arms. With each tool placement they were pushed out only to flow back in with each secure purchase. I reached far to my right and set my ice ax hard and deep, kicked hard with my right cramponed foot, then my left, removed my left tool, weighted the right—and then I popped. I managed to yell "Fall!" as I whipped across the face in a 15-foot arc.

I don't remember falling, only landing in jangle of axes, slings and hardware. A piece of ice the size of a car tire had broken off as I had placed lateral pressure on the ax.

I sorted myself from the mess, felt the knot of fear in my stomach, set myself in a more secure place … then everything became very peaceful. I relaxed; the rats of fear were no longer chewing my guts; my hunched shoulders dropped. I looked around. I could hear everything: the melt water now rushing under the ice, Alan calling up from 75 feet below, the creaking of the tools in the ice. I looked around. The canyon felt like the world was opened in front of me. I was relaxed, at peace as the thought came to me: So this is how it ends. This is what life looks like when we die. We will be where we are, and it will look as it is. Nothing more, nothing special. Sky, road, rock, people, just what happens to be there at that time. A bit disappointing for a young romantic. This is it.

I tiptoed back up to my previous high spot and gently crossed the last bit of ice to the rock to set up a rappel. On the ledge I found a very old "Z" piton with a tattered piece of webbing tied to it. Someone had done this before us.

WHO ARE WE?
Mapping Our Roles and Their Source

JOHN GOLDER
♂ SCMC

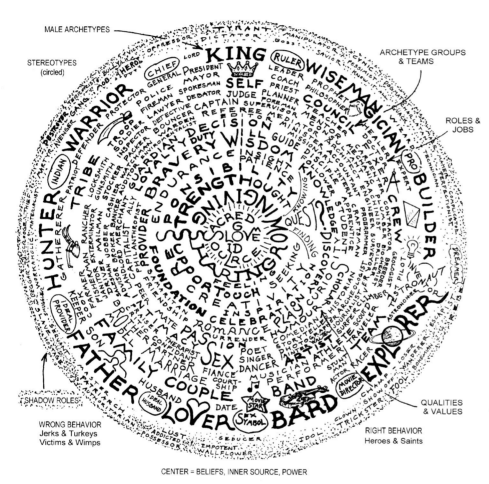

MALE ARCHETYPES

STEREOTYPES (circled)

ARCHETYPE GROUPS & TEAMS

ROLES & JOBS

SHADOW ROLES

WRONG BEHAVIOR
Jerks & Turkeys
Victims & Wimps

RIGHT BEHAVIOR
Heroes & Saints

QUALITIES & VALUES

CENTER = BELIEFS, INNER SOURCE, POWER

ORIGINAL DRAWING BY
THE AUTHOR

What is our role as self? Team? Mankind?

Above all else, it's what we *do* that defines us. Any meaningful change in our lives calls for a change in how we *act*, not just what we say or think.

Emerging from my mid-life crisis and searching the direction and source of my actions, I was introduced to R. Moore and D. Gillette's book *King, Warrior, Magician, Lover*. It quickly led me to C.J. Jung and other authors on male archetypes and men's identity. I was intrigued by these men's insights of male role models and set to work organizing and mapping the men I knew from childhood stories, songs, movies, and life itself.

THE ARCHETYPES

What emerged were nine to ten archetypal mature male role models of "right behavior" — King, Wiseman-Magician, Builder, Explorer, Bard, Lover, Father, Hunter, Warrior — and the anciently related archetypal groups from Self through Council, Crew, Team, Band, Couple, Family and Tribe. It is the effective teamwork of these groups that has created our civilization, but it is their breakdown, especially of those fundamental units of Family and Tribe, that threatens so much of our hard-won order.

WHAT IS OUR CENTER?

I'm particularly interested in the source beliefs that guide the behavior of these archetypes. What separates "the good guys" from "the bad guys"? Though they manifest in many different forms, what I found were three fundamental qualities central to the "Sacred Source": *giving*, *growing*, and *sharing*. Yet, there is a fourth essential and elusive element.

POWER, STRENGTH AND THE WILDMAN

Fundamental to men's actions is *power* and *strength*, which seem to rise from a deep amoral well of primeval survival energy. This is what Robert Bly describes as the "Wildman." Difficult to map, this energy flows through both "good" and "shadow" roles alike. Perhaps a wise Creator wanted to ensure our survival regardless of the path of our consciousness. Hence, the Wildman's irrepressibility can be difficult to control or predict. Like the Trickster, he can throw us off balance or save us from a tight situation. Understanding and using the power, flow and rhythms of our Wildman is one of the deepest challenges of the men's movement.

WHEN ARE WE BALANCED?

Picture the male symbol ♂ as a giant spinner in the center of the circle, an "identity compass" if you will, pointing in the direction of our self-esteem and action. I believe I am balanced when, by truly knowing the center and source of my action, I can turn to and access any role I need to on the circle. It is the centering (knowing my inner self and the beliefs that guide me) that allows me to swing wherever I need to on the circle without getting lost or entangled in the shadow roles and "wrong behavior" that lie outside.

WHAT ROLES HAVE WE REJECTED? SURRENDERED?

The focus and direction of our childhood role models has a tremendous effect on which roles we accept and reject as adults. Again using the

♂ spinner analogy, picture our children (or ourselves as children) riding the opposite end of the pointing arrow. As we point clearly in action through the focus of a Sacred Source, moving steadily and regularly between roles, our children get a clear picture of our actions and purpose in life. If we strictly limit our direction and roles, our children miss the full picture of male identity. If we swing too wildly about the circle, not only does the map become blurred and incomprehensible to them, but they are thrown off the opposite end by a very real reactionary force into the shadow roles that oppose our own excess.

WHAT ROLES ARE MOST VALUABLE TO US? OTHERS?

Western society bestows high financial rewards on our most effective leaders, wiseman-magicians, warriors and hunter-providers. Applying supply and demand logic, I think this points to a shortage of men with the underlying qualities of those archetypes; strength, bravery, will, wisdom, discipline and endurance. The enormous success of many public performers comes largely as a result of our fascination with their powerful portrayals of these archetypes.

To me the true value of these archetypes is that they serve as eternal examples to guide us back from the excess or shadow roles of the dysfunctional men in our lives (most often our fathers); oftentimes to the opposite side of the circle. From there we can look back across, through the perspective of our Sacred Source, and find the roles that heal our wounds and restore our wholeness.

WHO ARE WE? is an ongoing project. Contact John Golder, PO Box 854, Santa Cruz, CA 95061

Letter to Jack

PETER GRADJANSKY
Warthogs ♂ *EBNoM*

Dear Jack:

Thank you for the birthday card. It means a lot to me that you think of me through all these years though you hardly hear from me. Turning forty doesn't bother me in itself. And there are many wonderful changes in my life over these last years — including, above all, fatherhood — and, intermittently, the feeling that I have mastered certain skills and personal challenges, as well as the ability to earn a good living in a wholesome way: improving the world by creating beautiful, useful, valuable gardens for others, without compromising my values or my health or without doing very much that I don't enjoy.

Yet sometimes I wonder who I'm kidding. I sacrifice so much of what I used to think of as the essential "I" for every day that I live in getting and spending, and in taking on more and more responsibility. I'm left with not enough time for correspondence or spontaneous visits, or to be in nature, or read, see a movie, even rent a video, or work my own garden for pleasure. Fun? What's fun? I get barely the exercise that I need, and then only because my back problems would lay me up if I didn't. Really, anything I do, however enjoyable it could be in principle, can weigh on me like another task squeezed in, another scheduling problem.

Sometimes I lose the sense that I do things beautifully and completely, and to my own satisfaction, unless I'm being paid for it. And the unpaid chores of running my life and business pile up until I feel too often one short step ahead of chaos and one long step away from a state of pleasurably self-sustaining efficiency. From where I stand, I can see a long slippery slide into workaholism on the one hand, and a slippery slope toward lowered standards on the other — either way, to being less than fully *present* for my family, for my clients, for my gardens, for myself. So I feel myself on treacherous ground, where everything I care about is on the line.

Often it seems that I can't both carry on this "adult," bacon-earning, parenting lifestyle and hold on to the qualities that have made me "special." Could anybody love me if I were just ordinary? I think, "How can anybody manage financially, let alone ethically, wisely, with any kind of style or grace?" But obviously many people manage just fine, and with considerably fewer advantages than I have. I think the trick may be to have the courage and the good sense to be just ordinary, to relax enough to enjoy the freedom of that frame of mind, and to savor the feeling of common ground with all one's fellow humans (and other creatures).

It's time, I think, to review my expectations of myself: to realize that some expectations which once served me well may hold me back now; others may have crept into my assumptions recently, while my life has been changing so fast that I've just been scrambling to cope and adjust.

Take travel, for instance. Exotic, leisurely adventure has always been a yardstick for me for how my life was succeeding. Had I managed to keep myself sufficiently free and affluent and optimistic and energetic to relish strenuous trips to places most people wouldn't even think of going? Either faraway places with alien cultures or remote and challenging wilderness, or preferably some combination of the two? Lately, especially, I've looked to travel for reassurance that my youth was still with me. Yet at the same time, with bad back and a baby and a more limited timeframe, and with a wife with her own requirements, I've started to want my travel cushier and more secure. A contradiction, and an expensive one, as well as a recipe for less than total satisfaction, less than total relaxation. And in fact travel has come to seem more "necessary" than ever, as an accommodation to a higher-stress lifestyle; going away has come to seem like the only way I can relax and leave my busy-ness behind.

Increasingly, though, I'm finding that a trip can *create* stress, can in fact be an escape, in the negative sense, from the genuine centers of my life, which are currently home and work. Even if I return rested, I'm also further behind, especially if I take the long and strenuous journeys I've always loved.

These days, in fact, I take my heaviest responsibilities with me when I go, since I can't imagine leaving Hannah behind, and she's at an age where curiosity, desires, will and mobility have outpaced good sense, communication and survival skills. And for her, of course, there's plenty of adventure and novelty right here at home. Ironically, we can probably best help her to become confident and adventurous by sticking close to home, where she can take her early risks in a familiar and child-friendly environment.

So there are a lot of old judgments and habits I need to let go of, in order to do what's necessary here in this moment, and to accomplish my major current goals. If I can accept myself as "just" a family man, and a small businessman (gulp! this is not my self-image), perhaps I'll learn to relax without leaving the country — and then I won't need to earn so much, either, to support my travel habit.

While I'm at it, there are probably a bunch of more recent expectations in my head, about things like providing my family with *every* material comfort, which I ought to examine closely. If I can release myself from some of these self-imposed mental burdens, I may even manage to enjoy being "middle-aged," and keep myself young enough to go on adventuring — whether at home or far afield — for the rest of my life.

Maybe it's not over yet. When I think that, for all these twenty-some years I've really known you, you've been older than I am now, and that I still see you as a young and vital man, it gives me courage, and makes me hope I'll see you soon.

Love to you, Jack.

Peter

Harvest

MATT SORGENFREI
Quiet Riot ♂ SBNoM

A child sighs in his sleep
Fire crackles, leaps, and chuckles quietly
Rain falls like drums calling from the woods

Our world is more peaceful now
Because of the sweat and passion of our day
Sunlight saw our labor
The night sees us glow with honor

God expects us to laugh and to enjoy this life
That is the real secret of why grapes turn to wine
All within time, all by their own path
Age and sweeten and become spirit
Which brings a smile to the rest of us
Watching and waiting for our turn to shine

What I Would Have Done Anyway

ROGER CHAPMAN
Victory ♂ *SBNoM*

The biggest transformation in my life was subtle — so subtle that it could have happened without me even noticing it. Looking back on the Sterling Men's Weekend, and on myself, I see that the biggest transformation has been transforming thoughts and ideas into action.

So exactly what *did* happen after the Weekend? I was ready to leave my job. Morale was low and upper management seemed oblivious to problems in the company. I thought about what I wanted in a new job and what I would say at my first interview. It was then that I realized that, being a manager, they might ask me, "What did you do to fix any of the problems?" I decided to present the problems that I saw in the company to the president. Two other managers, also troubled, were to be in the meeting with me. But when it was time for the appointment, they backed out. I decided to present the problems with my proposed solutions. I figured that the worst thing that could happen would be to lose my job and I was ready to leave.

Two days after my presentation, the president called me into his office. He had decided to implement my proposals. Within a month I was presented with a 15% bonus for the Key Employee Award.

Was this a transformation? I always had these thoughts and ideas. Sure, I fought through my fear of confrontation and presented the ideas, but I would have done this anyway. Wouldn't I?

The next month I was thinking about going sky diving. I had always intended to go (for 15 years), but hadn't gotten around to it. I told a couple of people that I was going and they asked me to arrange a trip. The next month 6 of us were making our first jump!

Was this a transformation? I always had the intention of going. I was going to get around to it sooner or later.

I decided that I was going to run a marathon. Me, the slowest runner in my high school gym class. I always thought that this would be a major achievement. 26 miles did seem like a long way to run, but I always intended to run a marathon. As soon as I spoke the words to my team they wanted to see a plan. so that they could support me and keep me on track. Two members wanted me to run with them in the morning so that they could get into shape and lose some extra pounds. I completed the Marathon! 3:40, not bad for a first timer. Oh — did I forget to mention that I rode in the "Death Ride" the week before the marathon? 15,000 feet of climbing over 150 miles and 5 mountain passes in the Sierras.

Was this a transformation? I always had the intention of running a marathon.

What a coincidence. Within the first year of doing the Weekend, I had completed a lot of my "intentions" living life more fully. I was still the same man, but only different.

Besides all of the above things, I also went to New Orleans to see my college, Indiana University, win the NCAA basketball tournament. I also took a vacation trip to Maui. These were things that I always wanted to do, but I had never let myself do them in the past.

1987 was a very good year for me. I guess the biggest transformation was putting my dreams into action. Did I change? Well yes, and no. My intentions stayed the same, but the way that I showed up in the world, as a man of vision and as a man who lives his word, changed dramatically.

Now Makes The Past Right!

MIKE DRUKE
Tatonka ♂ *SBNoM*

Now makes the past right! Now makes the future welcome!

It has taken me forty years to reach those conclusions. I have arrived at a point in my life where I am comfortable with who I am now and what I have now.

For years I lived in a fog that clouded my perception of life. Guilt and shame from the past made my decisions. Fear of the future as an unknown, unwelcome uncertainty darkened my spirit. This fog cannot persist, this fear cannot remain if the Blessings of *NOW* are allowed to shine through.

Guilt about youthful, perhaps selfish, decisions convinced me that I had aborted my sons. This prevented me from celebrating my daughters. Shame about squandered potential deceived me into thinking that I was a failure. This scuttled my present productivity and accomplishments. Fear that I am unlovable and that all my relationships were destined to fail, prevented my from enjoying the woman of my dreams now.

When I fully appreciate the present reality, I know that I would never change anything in the past. Even the smallest change in my life might have prevented me from being at the celebration event on July 11, 1987 honoring recent Women's Weekend graduates, and therefore denied me the woman of my dreams. Any deviation in the past would have thwarted the unique sperms that differentiate Laurel and Heather. Therefore there is no doubt in my mind that I had to make the mistakes I did, suffer the pain that I did, hurt the people that I did, in order to be here now.

Getting By

ANDREW CREELY
Chi-Pachagon Express ♂ *SCMC*

It surprises us sometimes
when we get it right.
Like when I touched your hand and
said, "Your eyes just sparkle,"
or when we realize that we have
learned something about mountains or
poetry or love,
when for years we had thought we
were just getting by.

Perhaps we were — just getting by —
collecting the knowledge of sunsets and
the feel and smell of the coming winter,
side by side with the knowledge of
two plus two, take the next left,
this is the road,
this is home.

Reflections On Change

TOM BLOUNT

Victory ♂ *SBNoM*

As I sit here pondering what I should title this written work, I think back over the events of the last year and how they have changed my life.

It was March 28, 1992. A clear, cold Saturday morning in Oakland, California, and I was about to begin the event know as the Sterling Men's Weekend. My office mate and friend, Andrew, had finally convinced me to sign up for the Weekend after nearly two years of cajoling and (what I interpreted as) pestering. The way I saw it, my life was in order and I didn't need to go do some cult-like weekend event where I might be brainwashed. It was only after seeing the positive change in other men that I worked with and who had completed the Weekend, that I signed up. I also wanted to get Andrew off my case. In truth, I also hesitated because I was afraid. Afraid of wasting my money, afraid of doing something in front of other men that would embarrass me and afraid to really trust my friend.

The events of the weekend turned out to be pretty powerful. Although what I heard was not radically new to me, I had never seen the ideas presented put in such a light before. It made a hell of a lot of sense to me and I started to understand the purpose of the Men's Weekend; in short, to empower men to be the men they always wanted to be. I came out of the event fully energized but not quite ready to make any changes in my life (I still didn't think I needed any).

I joined a team of men right out of the Weekend. We got along pretty well and were pretty cocky having just completed a major event together. It was pretty cool being around a bunch of guys each week. Things were just getting interesting when a time conflict arose due to the illness of a close friend, and I had to break a commitment to a team meeting. This resulted in my involuntarily leaving the team after only two months. Standards had

been broken and the consequence was my leaving the team of men I had come to know.

I was pretty pissed off now and felt that being on a team wasn't worth the hassle. I thought to myself that it was just as well since now I had more time to pursue all the fun things that I didn't have time for before. Sure the Men's Weekend was still a worthwhile event, and I would recommend it, but the team part was only for those who had problems in their lives and I certainly didn't have any.

Early July, 1992. I was talking to John, another of my co-workers who had done the Men's Weekend, about being on teams. I related to him my previous team experience, and he explained to me that he was a member of a different organization of teams called the Nation of Men that was run a little differently from the organization tied to the Sterling Institute. He invited me to come to one of their meetings on a Saturday morning, and I committed myself to be there.

The meeting day came up and I forgot completely about it. Later that same day I realized that I had broken my commitment and I felt bad about it. I called John and asked what I could do to regain my honor. After some thought he asked me if I wanted to be on a team, and I replied yes, feeling that this would make him happy. John then told me to write down on paper what I wanted from a team and why. This turned out to be a real challenge. I hadn't a clue what I wanted or why but I wrote something about finding out about career goals and a purpose for my life. John added that he would get me in contact with teams that wanted to add new men.

I met most of the members of team Gunima around the beginning of August and seemed to connect with them. After a couple of meetings they were ready to vote on accepting me on the team as soon as all the members were present. I still vividly remember that fateful day when I tried to explain

M E T A M O R P H O S I S

what I wanted from a team. I dished out my best bullshit (it was so good I didn't even know it was bullshit) relating how my life was in pretty good shape and that I just wanted to improve on a good thing. I was hung up on making things look good to these new men I had just met. The team saw right through what I said and came back with some rather strong support. It was like putting up a mirror in front of me showing me what others saw: a sometimes arrogant, self-centered, son-of-a bitch.

Suddenly I felt like I was in an elevator descending faster and faster. I was scared for I had just caught a glimpse of one of my greatest barriers, and it unnerved me. My blood pressure rose and I started to sweat. At this point I had two choices: 1) put up the shields and go into denial or 2) open my eyes and accept what was being reflected back. Fortunately I chose the latter and, after putting my ego aside, I actually felt relieved. Being on a men's team is not about looking good but about being brutally honest and not afraid to trust in the men. I should have just admitted that I wasn't sure what I wanted and that I was there to find out.

By the following meeting Gunima had a new member. It took me awhile to get into the groove and understand the mechanics of how a team worked, but I started to see some of the little things that came with having other men in my life; that inquiry when I was down and things weren't going right; men asking how I was doing or how I felt. These were new experiences for me, and it felt good. Sure other friends would ask how I was doing but I was never sure if it was out of courtesy or genuine concern. I did get similar support from members of my family, but being thousands of miles away from them took away much of the immediacy. I really looked forward to the meeting each week where I could "recharge my batteries."

Gunima did some powerful things during the fall of 1992, many of which were centered around Star House, a home for wayward teenage boys. It felt awesome to be part of a basketball tournament and build a fence for the home then see how the boys responded positively to the presence of the men's teams in their lives.

My career situation was something to be desired. My vision of working in a world-class manufacturing environment was not consistent with reality. I felt like I was in a dead-end situation and had been thinking for some time about looking for a new job but kept putting it off. I brought it up to the team, and they supported me in better defining what was important and what kind of job I wanted. I felt empowered and within a few weeks had put together a resume and started on an active job search. I'm not sure how long it would have taken me to act if the team was not there to support me. Changing jobs or career can be scary but with the team behind me it was much less so. My life was really beginning to change.

The end of January 1993 saw the completion of team Gunima. Out of the ashes arose a new team, Victory, with all the men united in a common purpose: to empower men to experience and celebrate victory. Things were now starting to gel. The power and excitement level of the new team was building.

Other things started changing in my life as well. Due to a big upswing in demand for the products I am involved with at work, the job situation greatly improved. The opportunity to obtain a graduate degree at company expense came up, and I grabbed it and now am planning to start school in September. After much encouragement from a teammate, I stepped out to lead a joint picnic event between the Nation of Men and the Women's Community. Life was getting really exciting and it would not have happened without the support of the men on Victory.

Change is a wonderful thing. It can be scary and it can be exhilarating. But it is awesome when you have the support of men in your life. It is now one year almost to the day since I entered the Sterling Men's Weekend. I started out as a "doubting Thomas" and ended up as a believer in the process. I now realize that it was the start of something big, of unlocking the potential, of being the man I always wanted to be!

Hoka Hey

STEVE CROSS

Juggernaut ♂ SBNoM

Hoka Hey is the motto of Juggernaut, my team. In the Lakota Sioux dialect it means "Today is a good day to die." Not fatalist or anti-life, rather the most life-affirming statement I know. When life is lived so that every day becomes a good day to die, every day in that life is lived in appreciation, in wonderment, in beauty, in wide-eyed fascination.

When Hoka Hey sets the tone for the day, the aroma of the wild grasses growing in Palo Alto is noticed, not the traffic. When Hoka Hey sets the tone for the day, calls from friends are taken, gladly. Hoka Hey frees you to live in that wonderment, in that beauty, in that wide-eyed fascination with the magic of life.

Bottom Line

DAVID MEYER

Guest ♂ EBNoM

Human mission statement:
    Continual search
    Remembering and learning from
mistakes
    Living, from within, what we seek
    Time to move on (continue)

Personal philosophy of life:
    Give and forgive
    Impeccablility allows for happiness
    Enjoy life

Message to the World Number —?
    Think positively
    Act absolutely

Message to Mother Earth and to
    our mineral and organic siblings:
    We need each other!

Our Courtship Makes No Sound

DAVID RUBINE

Rod Warriors ♂ SCMC

Our courtship makes no sound.
An aching silence filled with promises,
it is the instant morning light skims
across still water,
the stinging cold that resolves night's hesitation
to accept morning,
a stone of passion skipping across dawn's lake.

Ours is that kind of courtship — a breathless
acceptance of stillness.

I anticipate you like a stone diverting a stream,
getting in your way just enough so you'll
brush lightly by.
I throw myself in your rivery path forcing you
to divert your course
and call that my love for you.

Rushing by you animate stones.
You teach stones to talk and they tell you
about their love.
You have learned to speak their silence.
Stones listen there.
Their love is terribly quiet; it is all they speak.

Teachers

DENNIS MANERS
Modern Warriors ♂ *EBNoM*

We all had teachers when we were growing up, starting the day we were born. Dad and Mom were the hardest teachers. Sometimes they taught you by yelling at you, and sometimes by taking their time.

The way I teach is the same way my father did. He taught the hard way. He said something one time and that was it. If you were eating with your mouth open at the table, you got asked one time; after that you got knocked off the table. It was a hard way to learn if you didn't get it the first time.

The same applied in school. If you messed up in class you went to the principal's office and got talked to one time. If that didn't work, ten swats with a wooden paddle.

It seemed that as I got older, the way my elders taught me began to change. My father took his time to teach me the things that were given to him by his father.

I have two boys; Jeremy is four and Joshua is five. When you have to teach a child how to walk, talk, eat and, yes, sleep, you know that there is a long way to go. Teaching them is a long-term commitment.

Times have changed the way children are taught. I believe in using whatever method you feel works best for you in teaching someone. As I get older I am learning different ways to teach through patience and understanding.

Thanks to all the men in my life who have taught me the difference.

Wild Man Inside

KIM WELCH
Chi-Pachagon Express ♂ *SCMC*

There is a Wild Man Inside.
It is born of the Great Wind.
The Breath of Spirit.

There is a wild Man inside.
A little boy in a way-big body.
There are wild men Outside.
Seeking, looking, playing Dumb.

There is a wild man Inside.
Who is born of softness / WOmb MAN.
Who is returning to the Light.
Who IS not separate but one.

There is a Wild Man inside.
To sign and howl to the Moon,
To Sister to Brother Sun,
The children of Father Sky and Mother Earth.

There is a wild Man inside.
And he calls to his brothers to Run.
With the Wind outside and in.

There is a Wild Man Inside,
Do you See? Do you Feel?
Yes, says the Wild Man Inside.

Changing Changing Gender

RAND WROBEL

Bucks ♂ SFT

ORIGINAL PEN & INK DRAWING BY
RODNEY CRAMER
Brujos ♂ EBNoM

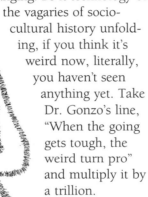

I. META-METAMORPHOSIS

MetaMorphosis is change. So Meta-MetaMorphosis is change of change. Like Toffler's thesis that the rate of change is increasingly increasing, so the nature of changing is changing. Be it technology or the vagaries of socio-cultural history unfolding, if you think it's weird now, literally, you haven't seen anything yet. Take Dr. Gonzo's line, "When the going gets tough, the weird turn pro" and multiply it by a trillion.

As a science-fiction buff and as a casual student of future studies, one can postulate a variety of futures with the proviso, "we don't know." I'd just like to state right off that (for myself):

Statements about the future have an indeterminate truth value.

Often some will assign probability values to future events. The above statement applies to the basis of any probability calculation. All this is to say, just guessing.

GENDER

So why talk about futures, especially when it involves weird (and maybe uncomfortable) topics like gender, manhood, womanhood, identity, and related issues? No real reason, besides this stuff will mess with your head, meaning serious reality checks are and will be applied to our cultures in the coming near future. One can simply note the phenomenon of the movie "The Crying Game" to demonstrate the magnitude of coming possibilities in the arena of sex.

With the movie "The Crying Game," a wide cultural shock occurred on a mass personal level: a mainstream worldwide audience's perception of gender was challenged. Has anybody shocked you with their actual gender? Are you ready for a future of such shocks? Gender identification has had its ups and downs, twists and turns in history. Speculations like "Mr. Mom" (a movie where a guy gets pregnant) and a limited number of transsexuals are about all the main American culture has been exposed to. Homo- and bi-sexual strains in mass media are on the rise. Lesbians, gays, pairs, groups, you name it, all becoming more open, more mainstream, more.

I'd postulate an increasing diversity of options in lifestyles, all more jammed together in an increasingly interconnected and privatized manner. Our "culture" could be described as a collection of lifestyles, ways of life, that coexist (to some degree of peace or violence). Where the friction happens between strains, sub-cultures clash, compete, compromise, even cooperate.

METAMORPHOSIS

A Bar

As an aside, I'll mention a funny (but perhaps not that uncommon) occurrence in a college bar in Emporia, Kansas. This bar had a rather nice upstairs, and it seems some number of women and gays had "discovered" a pleasant, quiet getaway bar. Well, sooner or later, a few straight guys discovered this gold mine of, excuse me, babes. So, the story goes, a precarious balance existed for some time of:

— gals getting away,

— gays getting together,

— a few straight guys chasing the babes,

and, of course, a few straight guys probably got propositioned by some gay guys ....

Now if this sort of goings-on is happening in Kansas, hello, it's happening big-time on the coasts, in the bigger cities. One might be surprised that the first prime-time lesbian kiss happened in recent history. Family values are a topic of intense debate now because the family, as one's lifestyle, is subjectively defined. You determine who your family is and how you treat them. So does everybody else.

Expectations

Aside from issues of age (that is, assuming a legal adult), what legal barriers do or will we have to **any** sexual persuasion? Homosexuality is still not free of legal estrangement. As the distinctions of gender fade and merge, how will we react?

Now, I'm speaking as a heterosexual male, "whatever anyone wants to do with someone else by their free will, it's their choice...." I'm happy with my girl friend. Just expect the whatever to get weird beyond your expectations.

II. Phenomenological Questions of Radical Change

Ever feel like nothing was the same? Ever feel each moment and tremble as you realized you couldn't keep up with the moments? Ever contemplated your sheer existence, a flick of the eye, a twitch of your hand? Did your head feel big or heavy?

Have you contemplated God and/or humbled and amazed yourself at the sheer unlikelihood of your own existence? Have you felt yourself at the bow of the great ship of your life, rushing through your time and space on this planet? Does the wind of life rustle your being?

III. Change As Loss: Team

Loss of team is a major loss. In one list:

Self
Mate
Family
Clan
Super-Clan

Desmon Morris, who wrote *The Naked Ape*, spoke of the unique human trait of more levels of allegiance than primates, who have more than most other species.

He supposed the Clan or "team" level to be the outgrowth of hunting groups, like predators in general, humans could hunt in packs successfully, so they did, and allegiances became important. As one of these sort of levels, particularly painful can be the rejection by the team. It is part of life, and it may hurt.

The Man in the Mirror

CHARLES MANCE
Bushwackers ♂ EBNoM

Who are men and what is our purpose as men? It has been suggested that men are not supposed to be happy. Is the purpose of life for the male in this society just to leave home at a young age, cohabit with a female, discipline his offspring, toil to provide for a family, and listen quietly to his spouse until he dies?

You must realize that whoever you are, your children will learn from that, and they will use it to shape their lives. If you are always angry, unhappy or disappointed, if you feel like a failure, you will operate your life from those feelings. Everyone and everything you come in contact with will reflect what you are experiencing. Your relationship with your woman, your children, friends, and fellow workers will be affected, damaged and possibly destroyed.

A man carrying baggage that produces feelings of despair, loss, loneliness or hopelessness should take a look in the mirror and ask himself, "What is my purpose?"

During my brief association with the Nation of Men, I've experienced many men giving much of themselves on behalf of others, and I've wondered to myself if these men are as together as they appear.

Any man whose life is not working for that man is in no position to help anyone else.

I believe that being a giver to those in need creates community; but a responsible giver begins by giving to the man in the mirror — himself.

Every man is entitled to enjoy love, success and happiness in his life. Creating these values in our lives should be our first concern. When a man feels fulfilled and operates out of that feeling, he can empower everyone he comes in contact with and positively transform everything he touches.

So, men, let each of us take a deep look inside himself and determine what is and is not working in his life. Then let us seek the support we need to take action to create the value that each man deserves in life.

Don't cheat the man in the mirror!

Health

TOM MCCOOK
Tatonka ♂ SBNoM

A state of optimum physical, mental and social well-being, not merely the absence of disease or infirmity

1. Breath! **It is essential to your vitality!** Three times per day take 10 breaths, breathing in for a count of 5, hold for 10, then exhale for 5. Say *aaaah*.

2. Minimum of 20-30 minutes of aerobic exercise, walking, biking, stair-climbing, hiking — 5 times per week.

3. Daily questions:
   - What can I do today to have more energy?
   - What can I learn today about health?
   - What can I do today to stay focused?
   - What in my life am I grateful for?
   - What can I do today to experience pleasure?

Here's to your health!

Descent

JOHN TIMOTHY
Brujos ♂ EBNoM

The late afternoon sunlight streams through the branches of the huge old California live oaks and glints off the raised brass lettering of the plaque. Forty-five years (almost to the day) after the dedication, these words — "Stockton swimmers" in particular — seem to me a little quaint.

But the people at that dedication ceremony in 1948 were all thinking of specific young men they had known, men who didn't come back from World War II ... beautiful young men who swam in the rivers and canals and small community swimming pools of the San Joaquin Valley. In 1948 it was easy to remember them, to see them in their bathing suits, with their strong young bodies, horsing around in the water with their kid sisters or with their girlfriends, laughing....

And those young men would never swim in this magnificent pool.

Now, in 1993, I'm pacing around the perimeter of the pool to take a rough measurement of its size. It seems to be double Olympic length, and wide in proportion. There are high and low boards, spacious changing rooms, a snack bar, a grassy area for sunbathing — all set in a large verdant city park, a typical feature of Central Valley towns.

In grief and remembrance the people of Stockton built this pool. Post-war America was mourning its dead.

Yet, with the Depression and the War finally behind them, people felt optimistic and confident and excited about the future and its possibilities. And the word "community" still made sense: we had come through hard times *together*, and we were going to build something *together* — a better, more prosperous society. Many of the great public buildings and other facilities we use today are a legacy of the spirit of the late '40's and early '50's.

I was a child of this era.

ORIGINAL PHOTOGRAPHS BY THE AUTHOR

METAMORPHOSIS

I grew up in Stockton. Today I'm back in my old home town for the day. I've finished my music gig at the shopping mall, and I've decided to check out some of my old stompin' grounds. Though I now live in Oakland, less than two hours west of Stockton, in the last 20 years I've only passed through Stockton without stopping, on my way to somewhere else. I have no friends or family here. But today I've decided to be a tourist in my own past.

My junior high buddies and I used to bicycle over to Oak Park Pool almost every night in the summer for night swimming. We were cool: doing can-openers off the high dive, looking down girls' bathing suit tops, and seeing how far we could swim underwater through the eerie blue glow of the big porthole-shaped underwater lights. Afterwards we'd buy nine tiny burgers for a dollar at White Castle and split them three ways. I was a smart-ass kid, self-confident and happy.

But times have changed. It's 1993 and, as I look through the chain-link fence, I see there's no water in Oak Park Pool. The gates are all locked. The diving boards have been taken down. The steel ladders have been unbolted and pulled up onto the deck.

What's going on? Maybe they haven't opened the pool yet for the season. But it's the middle of May and it's sure as hell hot enough for a swim!

More likely, Stockton (like everywhere else) is in a severe financial crunch, and the Parks & Rec budget has been slashed.

But more disturbing than the apparent financial crisis is the spiritual crisis: everywhere we have lost that shared feeling of confidence and optimism about the future. And we've lost the sense of community — the knowledge that we're all in this together.

That there is trash all over the ground in Oak Park is due more to a general lack of personal responsi-bility than to cutbacks in park staffing. People just don't care any more. Maybe that's why the dedica-tion on the plaque sounds so quaint: people so rarely make personal sacrifices for the community these days, and they are seldom honored when they do.

But even more disturbing to me than the collapse of *society* is the collapse of *my personal life*.

Over the past two years I've gradually awakened from a kind of dreamlike unconsciousness to find that a *crazy person* has been running my life for a long, long time!

I'm living in a warehouse full of cardboard boxes full of junk. My finances are a mess: my income isn't meeting expenses, I'm in debt, I haven't filed my income taxes since 1981. I have no health insurance and no retirement plan. There's no woman in my life. I haven't done anything creative in my music for years.

I seem to have lost my *story* about myself — you know, the story that we all keep telling ourselves about ourselves over and over and over? The story about our past — and about who we are now — and about where we're going? Well, I no longer believe my old story. And I haven't replaced it with a new one.

The other thing that has happened is that I no longer have the *myth of the unitary personality* — the myth that there is one person, "John Timothy." I still behave to the outer world most of the time as if that were the case, but it's just an act; I don't believe it myself. To myself I seem to be a collec-tion of many different personalities.

I'm dragging myself through life, trying to pay the bills and, out of a vestigial sense of personal responsibility, trying to keep the commitments I've already made. My enthusiasms are few and they are as transparent and meaningless to me as jellyfish. It's very hard to get out of bed in the morning.

METAMORPHOSIS

So this little tour of my old haunts on a warm afternoon in May, 1993, is not just a stroll down Memory Lane. I'm desperately searching for some kind of thread that will lead me out of the labyrinth.

I leave Oak Park and I drive three and a half miles back to the old neighborhood. Madison Elementary School is still there. It's painted institutional beige now instead of institutional green. The "temporary" portable classrooms from my school days are still in use.

Occasionally in my dreams I'm back at Madison, walking those halls, heading for the cafeteria or for Mrs. Lussenden's room.

California used to have the best public school system in the country; that, too, came from the optimism and commitment of the 40's and 50's. I was a beneficiary of that system. So how come I'm not successful?

It's Saturday; the classrooms are empty. But even if I walked in on a weekday, there's nobody here now who would remember me. When you're 20 or 25 you can go back and maybe find one of your old teachers. But not when you're 49.

I know Mrs. Lussenden is gone; she was old even in those days. Looking back, I see that she was the first of many people who tried to mentor me. She wanted to steer me towards writing. She turned me on to the adventure stories of Howard Pease. Pease was a local writer who was an adventurer himself, or at least a traveler; his books were about the sea and far-off mysterious ports of call.

I see now that many people in my life offered to mentor me in one way or another, or were available for that if I had only asked. But I was stubborn and headstrong and wanted to do it my own way, and I had no idea what "mentoring" was or why it might be valuable.

In elementary school I was a dreamer and a goof-off. But by high school I'd become interested in the challenge of academics. I had motivation, focus, discipline and self-confidence, and I brought all these to bear on my studies. Why have those qualities slipped away from me? What has happened to my character?

Two years ago I was forced to admit to myself that my 20-year mission of making a living as a musician had failed. And I had no one to blame but myself. I had not been able to apply the intelligence and discipline I once had to either my music or my finances.

I leave Madison School and head over to the old railroad bridge — really my main destination on this tour. This was the "magical" place that belonged just to us kids. Two or three times a week a local freight train lumbered over the bridge that spanned little Smith's Canal. The bridge itself was not a drawbridge but a *turning* bridge. The whole bridge rested on a big concrete block in the center of the canal; huge machinery beneath the bridge stood ready to pivot the whole bridge 90 degrees horizontally so that a sailboat, for example, could pass through on either side.

Of course, nobody had ever actually seen this happen. The rumor was that the bridge had been opened once after the war to allow a yacht to pass upstream a few blocks to American Legion Park, where Smith Canal ended in a little lake. This was supposedly part of a big celebration welcoming the troops home; an admiral or somebody important was on the yacht.

My buddies and I climbed all over this bridge. I don't know if we actually hunkered down among the huge gears as the freight train rumbled directly over us or if I just imagined doing that, imagined it so often and with such emotion that it has become as vivid as memory in my mind.

Near the bridge on both sides of the canal were wonderful open fields of tall, grassy weeds: a perfect setting for dirt-clod wars. The only building nearby was a big mysterious factory with its own rail siding.

As I drive into the area, I notice first that the fields are gone; the factory is gone; and even the railroad tracks are gone. Suburbia has taken over completely. Can I even reach the canal without trespassing through somebody's back yard?

I find one last section that has been subdivided, but where not every lot has been built on yet. I park the truck and walk through the remnants of the old fields and up the bank of the canal. But the old bridge is gone too! What a massive project that must have been to tear it out!

When the last open fields are paved over, when all the railroad tracks are torn up and all the bridges are torn down, where will the young boys go to get away from the adults and just be boys?

Hanging out together on the bridge, my friends and I talked about many things, including, as we got older, the mysteries of girls and sex. This was one arena in which I had some success during my 20's and 30's. But I haven't been able to find that permanent relationship that I have been open to for so long. Three years ago I met a woman I

wanted to marry, but she dumped me after a year. I still don't know why; all she could say was that she didn't feel comfortable. Maybe she just figured out how crazy I really am.

Walking back to the truck I marvel at how even something as massive as a railroad bridge can disappear without a trace....The wild anise, though, still grows among the tall grass of these fields; its licorice smell was part of the magic of this place. And the feisty crows still voice their outraged complaints when you enter their territory. Nature somehow hangs on.

Before I leave town I want to drive by the old house on Marine Avenue.

I had a happy childhood. Our family had a lot of fun together. And my parents loved each other and stayed faithfully married until they died. (I didn't appreciate how rare that was until I went away to college and talked to other young adults about *their* childhoods.)

At least, that's the story I've been telling myself all these years. But, as I said, I no longer believe my old story. Because if my childhood was so great, why in hell is my life so screwed up now? Isn't a happy childhood supposed to be the basis for a happy and successful adult life?

METAMORPHOSIS

My mom died in 1977; my dad in 1983. Just before he died, Dad confessed to my sister Ellen that he had carried on a long love affair with a woman during the late 1950's and early 1960's. Mom, I'm sure, never found out; she would have blown up! Apparently Dad kept this affair going even after this woman moved to San Jose. He only called it off when Mom first came down with breast cancer; whether his guilt finally got to him or he simply wanted to devote more of his energies to caring for Mom, I don't know.

It was just three years ago that my sister told me about Dad's confession to her years before. I was pissed off that Dad had told her but he hadn't told me! And I was angry at his recklessness: he had put the family at risk.... As I write these words I still feel the anger and confusion; I haven't yet made my peace with this revelation. This is part of the interior work that I still must do.

I turn the corner onto Marine Ave, and I am happy to see the old pine tree still standing in front of the house. High up in its branches my friends and I built a platform from scrap lumber. At night we would sit up there and throw water balloons down at unsuspecting passers-by.

The big redwood fence Dad and I built is gone, replaced by an ugly little chain-link fence. The house is basically the same, but it looks different; someone has added a new garage and most of the stuff in the yard is new.

But the fruit trees Dad planted are still there, and they look great! Of course, the people who live there now don't know that those trees were planted by Art Timothy; but I know.

And what else remains of Mom and Dad besides a few anonymous fruit trees and a lot of happy memories? What else did they leave behind?

Everything disappears: swimming pools, railroad bridges, teachers, parents, friends, lovers. Even the

wild anise and the crows and the fruit trees all will vanish.

Maybe the only thing that matters, ultimately, is a few good years of happy family life. My sisters, both younger than I, have successfully started families. I haven't. I received the blessings of growing up in a good family environment, but I didn't turn around and create that environment for anyone else. I didn't pass on the legacy.

Nor have I created any music or writing, or done any great civic deed, that will outlast me. It seems that I grew up in the best of times, with every advantage, and I've done nothing with what I was given. I haven't even planted a fruit tree.

✄

There's nothing left for me in Stockton. I could go on and on about stuff that happened and people I knew and emotions I felt, but so what? It's just another meaningless story. So I get back onto the freeway and drive off into the setting sun.

This afternoon spent in the old stompin' grounds has turned out to be just one more step in my descent into the ashes. There are many more steps to go; I know each one will be more difficult and painful than the last.

When I read Robert Bly's description of "ashes work" I thought, well, that's cool, you go through that process and come out the other side, and you feel better, more complete, more truly alive. But my sense now, in the middle of that process, is that it's a long spiral staircase downward, with no way out, or through, or back. And at the very bottom, the very last step of all … is death.

✗

AFTERWORD

On August 21 I drove back to Stockton with my friend Ron to take some photos for this article. Oak Park Pool was open, and people were enjoying it. The high dive was gone, though; it had been replaced by a second low board. The park itself was full of extended families and little communities picnicking peacefully together.

Out at the canal there was a guy fishing near the invisible railroad bridge. We drove around to the opposite bank to get a better picture. It was impossible to tell exactly where the bridge had been; the right-of-way on both sides

had been flattened out, and houses had been built in its place. The blackberry bushes, though, still grew thick along the canal bank; their fruit was warm and sweet, and the occasional path through them down to the water hinted that maybe local boys were still having adventures.

Back at Marine Street the pomegranates were beginning to ripen on the tree my father planted.

My descent continues. But I see the possibility, at least, that at the bottom lies not death, but simply a solid place to stand.

FATHERS

METAMORPHOSIS

METAMORPHOSIS

Ivan's Tools

JOHN OBERSTAR
NFL ♂ SBNoM

Okay, Dad, but not right now. Maybe later on today, O.K.?

✗

At home again. Not the one I bought in California, but the real McCoy, THE Original Home I grew up in with my parents. I love my folks, but if the truth be told, I don't like coming back to THE Home anymore. Mostly I convince them to come to my home by extolling the natural beauty of California. The only times I stop by THE Home now is when business takes me through O'Hare. Then I arrange to spend some time with the folks like I am doing now. I used to think what kept me away from here was that when I'm here I feel like a 15-year-old again. This time though, I think there's something else. Spending time with my parents in this house reminds me of the passage of time and our mortality.

✗

My father, Ivan Oberstar, has been an electrical contractor ever since he left Jugoslavia at the end of World War II. He is an Old World craftsman of wire and conduit.

When I was growing up, Dad used to take me or my brother on jobs with him. It was his way of making us earn our keep as well as teaching us his craft. Generally our responsibilities consisted of pulling wire, fetching tools, holding things, or cleaning up. There was always this edge to working on jobs with Dad. One of his quirks was keeping lines live that we were working on. I swear to you that he is impervious to 120 VAC. It was always humorous to me and my brother to watch people's reactions when they would see Dad grab live 120 volt lines without even twitching. What they didn't know was that he has such thick, callused skin on his hands that there

was no way current was going to flow through his body. My younger brother loved working with my Dad. Today he's a lot better at home wiring than I am, because he took my Dad's lessons seriously. For me, the stuff was a drag. I preferred staying at home and reading books. Back then there were only three reasons to work with Dad: it was always interesting to see other people's homes and how they lived; he needed the help because sometimes the jobs were tough, and he came home dog-tired; and, it was a chance to be with my father who wasn't around home much because he was always working to provide for the family.

I remember being eleven and waiting with my Dad at City Hall for a permit when a city inspector pulled me aside. "Your father's a good man, boy. We stopped inspecting his jobs years ago. No reason to. Sometimes I tell the guys who fail inspections to go out and take a look at one of your Dad's jobs. You listen to your father and you'll be fine."

I remember being proud of my Dad's work and what he did from that day on. Ivan did great work, and he charged at or below going rates. If you were having "money problems" he'd take that into account too, despite the fact that we weren't always in the best of shape. A job was never done until everything was cleaned up. When he left a job the only thing left for you to do was throw the switch. You'd never know that he had been on the job except that things now worked, or you had new lights or outlets around. He was the kind of man you kept using once he did a job for you, no doubt about it. Ivan is an honest man with a lot of integrity about his life and his work. My father retired from the business eight years ago. Nowadays he just does work for friends, widows, or churches.

Mark —
Thanks for your friendship & support!
John

Like I said, here I am at THE Home for the weekend. It's Friday evening, and I'm here till Sunday evening. I'm having dinner with Mom and Dad.

> *John, I have so many tools at home I can't use no more. Why don't you choose some and I send to you. I don't need some many no more.*

> *Yea, sure Dad, but not right now. OK?*

⚘

Saturday noon, lunch ...

> *John, let's go down in beesement and look at tools?*

> *Ah, Dad. Can't we do that later? Let's watch Star Trek right now.*

⚘

Saturday evening, dinner with parents and brother ...

> *Why don't you two go downstairs and look at tools? I tink you need more dan me. Let's go.*

> *Come on, Dad. What am I going to do with that huge tripod mounted pipe bender or that monster cable cutter of yours? Give me a break.*

⚘

Sunday morning ...

> *John, ve go downstairs now and choose tools. No more bullsheet. Are my tools so junk?*

> *OK, OK. Let's go.*

So here we are in the "beesement." My Dad pulls open a drawer filled with a lifetime's worth of wire cutters, crimpers, screwdrivers, and you-name-it-it's-in-the-drawer. He picks up this huge pair of channel locks and looks at them carefully.

> *Boy, dees bring back old jobs, huh John? Remember when ...*

I can't answer. You see I left as soon as this started. Luckily I was standing behind my father so he couldn't see the tears starting to come out of my eyes. I never did answer him. I turned my back and went for a long walk. I finally figured out why I could never get myself down into the beesement even though I love tools. I told him why when I got back. His handing out those tools reminds me he's near the end of his life and so is my Mom.

> *Ah sheet, John. It just mean I have tree pair of channel lock, and I only need one! Mom and me is still healthy.*

⚘

The tools my father sent me are special. They sit in my best toolbox among my best tools, although they are neither the newest or the best. My father's sweat and dreams have seeped into these tools. When I hold them I remember my father. I'm saving these tools to pass onto my child so he'll feel his grandfather's strength and life.

Thanks, Dad

BOB KEYES
Modern Warriors ♂ EBNoM

Early in the morning of my twenty-first birthday my father came to me and took my hand and said, "Come with me." He took me down the stairs and out in front of the house.

He turned to me and said, "Everything you see around you is the world. I'm giving this to you today. Do with it what you will. It is yours, Bobby."

At this point he turned and walked back into the house, leaving me standing there totally confused.

Every year on August fourth, the very first thing I think of is his gift to me. The understanding and value becomes clearer and stronger every year.

Again, thanks, Dad.

Where All The Dicks Hang Out

MICHAEL SIEGLER
Tatonka ♂ SBNoM

Ten years old,
In a public bathroom with Daddy,
Struggling to take a leak 'cause
    all the urinals are too high.

All older men ...
No kids my age ...
Nobody talking ...
Everyone looks straight ahead
    in that quiet men's room trance ...
Just the sound of guys pissing.
Daddy leans over to me and quietly whispers,
    "I hear this is where all the Dicks
    hang out."
It cuts through all the tension.
I start laughing because I never expected
    him to be so cool,
Especially with all the stuffed shirts
    staring ahead at the bathroom tiles.

Twenty-five years later,
In a public bathroom without the old man,
Same stuffed shirt tension
    staring at bathroom tiles,
I laugh to myself
    'cause no one else will get the joke.
Dad, you were right:
This IS where all the Dicks hang out.
I love you.

Life Metamorphosis: From Drunk to Sober to Dead

PETER ALEXANDER
Warthogs ♂ EBNoM

Ever since he was fourteen, my stepfather Bill was the center of attention. First as a Vaudeville performer, then as a radio personality, jazz band leader, and outside salesman. He thrived on entertaining people in all facets of his life, and he had more friends than a new lottery millionaire.

Unfortunately, Bill and I never got along when I was a kid. He was a heavy drinker and I held very little respect for him. I remember Bill would call me a "punk." I would retaliate by calling him a "drunk." And his reply would be, "Yeah, but I'll be sober in the morning — you'll still be a punk."

This scenario went on until I was 21 years old. That year Bill decided to quit drinking rather than risk going to jail for numerous D.U.I.'s. The following two years we finally had a good relationship. We would share jokes, sales strategies, marketing and advertising ideas, and spend time together just hanging out. It was the closest I had ever come to having a Dad, since my real father was always too busy to take an interest in my life.

When I was 23 years old, Bill started having problems with his legs. He had trouble walking, standing, sitting up, and even lying down with his legs straight. Three failed back surgeries later, he became a paraplegic. Now he was bedridden and spent his days watching TV, reading and sleeping.

Needless to say, Bill's attitude towards life began to deteriorate. No more jazz sets, no more sales calls, and no more friends. The people who used to call him to hear the latest joke or story were only hearing how much he hurt or what he saw on TV that day. In a span of 18 months he had completely lost the ability to be the center of attention, something he had mastered for 35 years.

Things at home were not much better. He became angry and nasty with my Mom, depressed and difficult with me, and uninterested in helping himself lead a productive life as a handicapped citizen. (It was easier for him to make excuses why he couldn't do something rather than trying.) My mother and I kept sharp objects and his medicine (except his next dose) away from him for fear he would try to commit suicide. I had the misfortune of watching a person who wanted to die slowly taking his partner (my Mom) with him mentally. Even Bill's own kids wouldn't stop by or call him because they also were tired of hearing about his pain. Five days after Thanksgiving in 1988, I came home and found Bill dead. He had managed to hang himself with an arm exercising rope his son had given him months before.

For several months I was very angry over what he had done, but with the support of my friends I came to accept this for what it was: a tragedy. Bill was a smart-ass when he was drunk, a wonderful man when he was sober and healthy, and a tragic man when his health declined and he decided to give up wanting to live. He was bright, intelligent, funny, and always ready to entertain. Now that he is gone I cannot help but feel cheated that we didn't get to spend more time together when he was healthy. On the other hand, I consider myself lucky for getting a couple of years of friendship with Bill during the 15 years I knew him.

Thanks for those two years, Bill. I will always remember you in my heart as my Dad.

Thanks

JOHN COLLIER GROUNDS
Grunt 'n' Pokers ♂ SBNoM

*When you were getting ready to leave you said
by way of good-bye, "I'll catch up with you later,"
and I was too numb with grief to understand.
Your passing was a terrible loss to me
until those thoughts
gradually faded in the rush of day to day.
Now there's a new sense of a greater terrain,
a larger viewpoint.
I want you to know that no one was ever a
better father who
lived with more love and joy.
A terrible loss, and a great gift, and when
my time is come
I look forward to a noisy reunion.*

The Most

FRANK JOHNSON
Quiet Riot ♂ SBNoM

*You're the one I love the most,
You've taught me tree's shape and size,
You've taught me people and to look in their eyes.
You've given me the love I can pass down to my family,
The way you did to me.
You make getting older fun and not work,
But what really hurt me deep down inside,
Is I have to say good-bye to the one I love the most,
You.*

Honor Thy Father

Jordan S. Gruber

Tatonka ♂ *SBNoM*

"So tell us," said one of the men, "why do you want to be on this team?"

"Well," I said, "when I imagine the most positive possible me in the far future, and then I imagine that me looking back over his past, it's clear that being on a men's team is something that he would have — I mean that I would have — done."

The big, burly, gruff guy in the corner grunted. "That," he said, "is utter bullshit. You've been yanking our chain for fifteen minutes now and I don't have any idea what you're talking about. Why don't you try getting out of your head?"

I was stunned. I knew things in my "interview" weren't going that well, but I hadn't realized that they were going that badly either.

Gently but firmly, one of the oldest — and certainly one of the biggest — men in that room full of imposing, unfriendly, men said, "Tell us about your father."

So I began to tell them about my father. I told them how he didn't teach me how to fight or play sports, how he never really taught me to be a man, and how I had always felt it was necessary to turn to my mother and not my father when something important was up.

When I was done, the same man nodded his head somewhat sadly and said, "You know, I asked you to tell us about your father, and all you could do was tell us bad things. I want you to try again, but this time I want you to tell the story of your father in a way that honors him."

So I told them the *other* story of my father: how despite growing up in abject poverty without any decent male role models he had provided for his family for forty-five years without fail; how he had sold furniture on credit as a "peddler" in some of the most dangerous parts of New York City because that was the only opportunity open to him; how, as a peddler, he had been held up at knifepoint and gunpoint seven or eight times and how, despite his frequent nightmares, he had kept going; how he had decided to become a stockbroker at the age of forty-four even though he only had a commercial high school education, and how it was nearly unheard of for someone who hadn't gone to college to be allowed to take the broker's test and then pass it; and finally, how my father had maintained a strong, committed, and loving relationship with my mother.

A few weeks later, when I had been provisionally accepted on the team, I told some of the men that when I had told them the story of my father I had felt tears welling up in my eyes. They asked me why I didn't allow those tears, and I said I didn't know why. But I do know that this time, while I wrote this page for my father, I let those tears flow freely, because I know now that my father is a great man, that it is my immense privilege to be his son, and that I will love him and honor him forever.

It's All Washed Clean

RAY GEORGE

Men of Honor ♂ SCMC

I could live to be a hundred or I could die right now
Daddy, Daddy, it's all washed clean
I could live to be a hundred or I could die right now
Daddy, Daddy, it's all washed clean

Once upon a long time
I was yours, and you were mine
Your only son, fulfilled a dream
Daddy, Daddy, it's all washed clean

We used to play in the back yard
I'd catch the ball, you'd throw it hard
I learned so much about self-esteem
Daddy, Daddy, it's all washed clean

All washed clean, all washed clean,
Daddy, Daddy, it's all washed clean

On my way from child to man
I felt the need to take my stand
What was to come, was unforseen
Daddy, Daddy, it's all washed clean

Now the tears can flow, I can let them go
I can embrace the man that was my dad

I could live to be a hundred or I could die right now
Daddy, Daddy, it's all washed clean

My dear old dad has gone to rest
He did his job, he passed his test
He left a boy just turned seventeen
Daddy, Daddy, it's all washed clean
Though years have passed, memories stay keen
Daddy, Daddy, it's all washed clean

I could live to be a hundred or I could die right now
Daddy, Daddy, it's all washed clean

When Dad's Time Comes

PAUL TYNER

More Power! ♂ SBNoM

I closed my father's eyes for him at 2:30 a.m. while the hospital staff outside continued in routine fashion. My father had just breathed his last breath, but only my sister and I knew it so far. We agreed to a quiet moment of reflection before notifying anyone; soon of course, everyone would know. His position as a judge in Washington County, Oregon caused the local reporters to pester us for details of his sudden hospitalization. The family and hospital staff kept mum. None of us even questioned the need for this to be a private situation.

Mary and I had continued our family's 24 hour vigil into that early morning shift. Pretty much the whole family had come to understand that his time had come.

Those of us up from the Bay Area, plus the wives and new grandson, had spent the last week in somber reflection on what we were seeing. Cancer can be vicious, and also peaceful at the same time. At 63, most would say he should have had more time. But then, they probably didn't know him. He was rich with his innermost satisfactions. Even while the creeping cancerous cells of a rare leukemia finally struck with a sudden fatal attack, the actual unfolding of events seemed to belie that.

What actually seemed so unlikely, so unfathomable, so *unplanned* was in fact, as it turns out, *well provided for.* You see my father had made significant arrangements beforehand. For example, the house was paid off, the IRA's we didn't really know much about were mature and substantial, the pension arrangements set so that the security of his wife was assured.

The eldest son John, also an attorney, scrambled and got a legal order signed in time to provide

METAMORPHOSIS

disability retirement with increased benefits. Somewhat later we found out that the page in the policy book had post-it notes attached to this special section ... showing and illuminating the way. Why did he work so hard, for so long, preferring his courtroom and legal routine rather than early retirement and perhaps some new hobbies?

If there was any singular joy in this very private man's life, it was the family he and his wife Martha raised. In an era where it was unfashionable to sacrifice, or to have large families, these two parents raised eight healthy well-adjusted happy children in an old-fashioned Catholic belief structure, and to excellent, still unfolding results.

Sometimes I try to imagine my dad in a golf cart — tooling around the course probably with some buddies or something — but the image dissolves as fast I try to conjure it up. Golf wasn't his bag. Retirement wasn't a driving ambition. What he loved were simpler things. Like his own family. Like sitting at the dinner table and talking about current events or other news with the family.

There was a time when I thought it was strange that not too many people came to visit, that perhaps my dad just wasn't very social. Then I went to the Requiem Mass. The church was packed! The courthouse in Hillsborough must have been stopped still. Because there in the church pews of my old boyhood town church sat lawyers and judges and parishioners and family and friends that seemed to jam pack the isles. Everyone pressed up to see us, and offer whatever condolences that they felt they could.

There was a special luncheon treat afterwards which was set up by the Ladies in the Church. It was so splendid — there was talk about a whole-sale conversion of the entire County Courthouse community. Friends I hadn't seen in years showed up. And somewhere amid the grief and sorrow, we, the close family of Judge John J. Tyner, Jr., learned an important truth about life and how to fully live it when we found and witnessed the love and appreciation instilled in so many people for the impact that John Tyner made.

I owe my existence to John J. Tyner. As his second son, there are many things which I would like to do to acknowledge him. I think back occasionally to the times when I could have done more to show how much I loved and appreciated him. But you never get another chance for some things.

What that leaves me with, and what that leaves the rest of my very proud family with, is a standard of sacrifice, caring and purpose that will always be remembered, will always be profound, and will always be damn hard to match. Forget the golf carts, forget the cruise ships, part of what I have learned is about the things which really matter. I hope my purpose will have been as clear, when in fact my time comes.

ORIGINAL PHOTOGRAPH BY THE AUTHOR

If Only I Could Have Spoken These Things When He Was Alive

WILL SCOTT

Brujos ♂ EBNoM

In April 1987 my father died following a lengthy illness, three months after my mother passed away. When he was hospitalized for the last time my sisters didn't know he was going to die so they didn't contact me. Consequently I wasn't there when he left. I knew when it happened though. In the middle of a session with a client it suddenly felt as if a vice grip was crushing my head. The word "dad" went through my mind and I knew he had died. When I called one of my sisters a little while later she said they had just heard from the hospital a few minutes before.

My father and I had a lot of unfinished business. He never revealed himself emotionally to me. I found out more about him from some of the people that came to the funeral than I had known through 44 years of being his son. After his death I would occasionally find myself in a place of deep sadness over the relationship we never had and things I never said. In the fall of '91, as I was healing from a long relationship that had ended that previous spring, my creative nature burst open and for the next six months poems came easily and frequently. I'd never written a poem before. These poems are arranged in the order they were written and reveal the healing transformation of my relationship with my father that emerged through this work.

✍

YOU LEFT TOO SOON
(NOVEMBER, 1991)

Bare branches of the Maple reach out over you,
the cold earth blanketed with golds, reds and browns of autumn.
You are at peace as never before.
I stand in the cold breeze with a hunger unsatisfied.

You left too soon for me,
then again, forever may not have been long enough
for our hearts to truly meet
and find who each the other was.

You and I walked different paths
careful not to reveal the souls inside.
In spite of that you taught me much.
Many things about you I truly loved.

Your beauty and truth lay hid
beneath your fear of living and dying
My love for you hid beneath an anger
that even I could not push through.

Desperately I wanted you there for me.
I wanted to pull you to me
the fear that you wouldn't be there was too much.
I wanted you to know my truth and beauty.

In your death your final lesson comes.
Never again to hide from who I am.
Wild heart scream and shout.
Wild heart run free.

METAMORPHOSIS

WILD HEART #2
(DECEMBER, 1991)

Icicles melt unresisting in the sun.
Willingly they feed the next generation
of earth's great bounty.
The never-ending cycle continues relentlessly.

Hopes and dreams litter the path I've tread.
Or were they mere fantasies
now turned to dust and bones.
Wild heart wake up and see your light.

Winter wind whistles through the cracks
of the long forgotten shack.
The cloudy day occludes the sun
but not its rays of violet spectrum.

I avoid the eye of my own success
choosing too often the ugly and dead instead.
Marvel do I at the ease of disappointing myself.
Wild heart rise up and stoke your fire.

Footprints in the virgin snow
leading nowhere in particular.
The graves are covered white and pure,
there is no one there that shouldn't be.

My heart knows more than it is told.
Life is too short not to believe.
Wild heart throbbing with desire,
run and dance for all you are worth.

The breeze so fragrant with nature's sweetness
rustles the newborn leaves of spring.
Nature exposes itself to all
knowing it has no reason to hide.

New life appears in every turn
even to those that don't believe.
Wild heart take heed these lessons of the earth.
From yesterday's decay the mighty Oak does grow.

THREE SHORT POEMS TO MY FATHER
(JANUARY, 1992)

You were a rare gift in my life
One I could not ever seem
to unwrap
to see clearly and admire
Holding you up to the light
even now
who you were
is still opaque in so many ways

It's taken so many years
to realize
you loved me very deeply.
My love for you
I never could show,
either.

All those many years
of senseless battles
that kept us apart
fills me so
that even now
I cannot think of you
without the door to my heart
wanting to slam shut
out of habit.

I Miss You Dad
(March, 1992)

You've been on my mind again.
Those many heartfelt statements
I wanted to say
but couldn't
Keep bubbling up in my thoughts

I am so sorry
you are not in my life
anymore.
Now that I'm finally able
to let my heart speak more bravely.
My soul aches
at all the distance I kept
between us.

I wanted to be there
for you
when the time came
to leave us.
I'm sorry I didn't make it.

I love you more
than I allowed myself to believe
for so many years.
I'm glad I know that now.

FATHER'S SON
(April, 1992)

My father's son
I AM
and
I rejoice in that.

I am
my father's son
and
I rejoice in that.

His blood and energy
meld with mine.

Much that
he also
offered me
I rejected
offhand.

It was only myself I sought.

It took the passage
of much time
for me
to realize it
was only my Self
he offered.

Original drawing by
Bill McConnell
Phoenix ♂ SBNoM

FATHERS

My Beijing Healing

ROGER BEHNKEN
Wailers ♂ EBNoM

Growing up, my relationship with my father was mostly about his anger toward me, as I was the black sheep of our family. Many times his anger scared me and left me with deep feelings of pain and guilt. In my twenties, as I matured, our relationship changed and began to grow. He started to respect who I was as a person. But before this developing relationship could become a friendship my dad died suddenly of a heart attack. He was only 56 and I was 25.

In August of 1985 I was traveling with a friend through China. One night I had the most powerful dream of my life. We were staying at the Guanghua Hotel in Beijing and late at night they turn off all the lights, so I stumbled down three dark flights of stairs and out into the street to find enough light to record my experience:

As the dream opens my father is returning home after being gone for three years because of some sort of illness. He is still somewhat disabled, traumatized and unable to handle much stress of any sort. My older brother and I had been doing our best to maintain his house, but it had fallen into some disrepair because we were both working full-time. He asks us to take him around and show him what needs to be done.

The main problem is some large tree branches which are jammed together between a couple of trees and perilously suspended above the house. They are dangerously close to falling onto the roof, doing great damage. We enter our cellar to look for the proper rope and tackle needed to prevent this impending disaster.

My brother and I cannot find the equipment and tools he needs, and my father angrily blames me for their disappearance. A loud, angry argument ensues. I urge my brother to stop my father from harassing me because I am ready to deck him, but my brother is helpless to intervene.

My dad and I are now screaming at each other, and I'm ready to throw a punch but I keep walking away, only to be drawn into the argument again by his accusations. Finally, I grab hold of him and tell him I only want to love him and ask him why we can't work this out? We hold onto each other and weep. He tells me that he loves me.

Since that dream, all of the anger I felt about my father and about his death in 1969 has left me. My memories of him now are warm and compassionate, giving me pleasure. I feel our relationship is finally healed.

The Man I've Always Wanted To Be

DONALD VICTOR MORGAN
Tatonka ♂ SBNoM

For years I blamed my father for my shortcomings. If only he had been less distant when I needed someone. If only he had taught me more when I missed opportunities in my life. If only he had showed me really how to be a man. In the many times when I looked for a scapegoat to blame, it was the impassive face of my father that shined brightly on the screen of my mind. There I acted out a drama in which I somehow received the magical wisdom, awareness, and strength that would have empowered me to avoid the hard times of my life.

Last Spring I went backpacking with my father in the Sierras. It was the first time as an adult that I have spent time alone with him. For four days we were together eating, sleeping, and exploring with no distractions. We were immersed in the wilderness. Although not many words were expressed between us, something deep and powerful was born.

I remembered how much I had respected and adored my father as a young boy — how through my seven-year-old eyes, he was almost a God. As I grew up, however, my father became to me a distant, unfeeling, unfocused presence in my life. I believed that he did not give me what I really needed.

Justin Sterling says that the role of the father is to make the life of his growing son as challenging as possible so that he is best prepared to be successful in a competitive world. I realized that my father had given me all he had and more. Every time I blamed my father for my own shortcomings, I was really only affirming my own lack of self-esteem and unwillingness to grow.

I saw clearly in the mountains with my father that the life he has led, and the life he has now, is very, very good. He is a man who has done, and is doing, all that he wants to do in his life. He spends his time now traveling, doing volunteer work, and being with my mother. He is happy.

During our time in the mountains I began to see my father with new eyes. I now know that in a strange, timeless, and familiar way, I am now seeing my father again as the man I have always wanted to be.

Dad

STEVE CROSS
Juggernaut ♂ SBNoM

My father once seemed to know everything there was to know. He knew everyone important and everything anyone needed to know. Then, as years passed, he must have forgotten all this stuff. He didn't seem to know much of anything anymore and what he knew was all wrong. If it wasn't wrong, why would we have argued about it so much? Then, years passed, and he seemed to know other stuff. He doesn't know everyone and everything anymore, rather, he seems to know a way to live. He has a quiet dignity he never showed before. A strength of humor and character that match the dignity. I used to think I knew who he was. That limited both of us. The man I now see is so much greater than who I thought I knew. Not perfect. Just a guy. Lots of faults. Hell of a great dad. Love him.

A Father's Miracle

JOHN A. LOZANO
Phoenix ♂ SBNoM

It was a Saturday afternoon. I was doing household chores and looking forward to the South Bay Nation of Men's "Shake Me 2 Baby" event that I would be attending that evening with my team. All the major appliances were running — washer, dryer, dishwasher, pool pump. I turned on the oven to cook some chicken legs for the event, and suddenly the electricity went out throughout the house.

I knew it had to be the master breaker switch. I turned off the appliances and went to the breaker panel. Sure enough, the breaker had tripped. I tried to switch it back on, but surprise! The switch wouldn't latch. I played with it for 10 minutes and could not get it to latch.

I went to the local hardware store. Orchard Supply was my first stop. They didn't have the 100 amp breaker style I needed. I then went to Home Depot (my favorite store), Ace Hardware, and McIvores Hardware. None had the proper breaker. Mr. McIvore recommended a place in Milpitas, but of course, they are not open on weekends. I drove back home and frantically searched the yellow pages and made calls. No one who was open had the part I needed. I started to think about moving the contents of the freezer to my parents' house.

Running out of alternatives, I called some team members for advice and to let them know I wasn't going to make it to "Shake Me 2 Baby." This was a major disappointment.

Finally, I called my father and asked him if he could think of anything I could do. He said "Hit the switch with a hammer."

I said "What?"

"Yeah," he said. "Tap it with a hammer or the end of a screwdriver." I was thinking he was a little nuts. But what other choices did I have?

I went to the breaker box with my hammer. I tapped it a few times ... pushed the switch ... and lo and behold, it worked!

By the way, I did make it to "Shake Me 2 Baby!" To me this was a miracle. Thanks Dad!

Dear Dad

SCOTT LUCAS
Sperm Whales ♂ SCMC

I miss you and wish I would
have come home in time.
Time to be with you eye to eye
and shoulder to shoulder.
Instead, all I have are wishful
conversations.
But this I know to be true,
you love me and I love you.

I Will Remember

ORIGINAL PHOTOGRAPH
BY THE AUTHOR

JOHN LASERSOHN
Grunt 'n' Pokers ♂ SBNoM

The lake grieves with me for you, Dad.
Frosted in wind-whipped
Rage and other times solemn,
Limp in winter doldrums.
Comforted by ducks tracing wisps
Onto dark reflections of silent firs
and shimmering pieces of the moon.

The lake rejoices for you, Dad.
Drifting patches defy wind and rain with
Triumphant shining, a current or maybe
a spirit in the water, telling me that
you are felt, you are loved and remembered.

(Read aloud at his Dad's memorial service)

Two Halves Don't Make a Whole

RON TUGENDER
Victory ♂ SBNoM

This one gave me life
His blood flows through my veins
He left his indelible mark on me
And then disappeared

This one rescued me from need and want
He took me as his own
He gave me all within his power
Except himself

One imprinted me and then abandoned
One sustained me and never bonded
Two halves don't quite make a whole
That part is up to me

METAMORPHOSIS

FAMILY OF MEN

METAMORPHOSIS

Mac & Me

GARNER MCALEER
♂ EBNoM

This photo was taken in 1957 on Mobile Bay. That's me on the left; brother Buck (Thomas Joseph McAleer, Jr.) in the middle; and cousin Mac (Joseph Arends McAleer, Jr.) on the right. Mac and I were inseparable in those days. We were seven and eight when this picture was taken; he is one year and a month younger than I am. The wharf behind us leads back to our house on a hill under moss-covered oak and pine.

Mac's family and mine lived there together during the summers. Tom and Joe, the dads, were chronologically the closest in a family of five boys and five girls. They have remained great pals throughout their lives. Joe, Mac's father, opened a donut shop in 1957, shortly after this photo was taken. The business expanded and it's now a big chain of donut shops all over the East Coast. Joe Senior retired a few years ago, and Mac took over the business.

Mac and I used to deliver donuts together when we were in high school. Before that we made donuts in one of the little donut shops. He and I went to an all-boys Catholic high school, the same one our dads went to.

Our families were close. There were a lot of kids. I had 60 first cousins just on the McAleer side of the family. Family, and all those associations, was part of what I was getting away from when I finally left Mobile. I was born and raised there; I even went to college there. It was my entire small world.

After college I went into the army for a couple of years, then I came out here to San Francisco. Mac and I didn't have much contact after that. He has three teenage kids now, and I've never met them; that's how much of a disconnect there was.

What's evoked for me when I see this photo or when I talk to him on the phone is that Mac has a real fun out-look on the world and his place in it. And I think he sees me the same way.

These days he's way up there in the serious corporate world of bank presidents and attorneys and so forth. He started telephoning me a couple of years ago when he was feeling deeply immersed in that world. He wanted to get back in touch with the spirit we shared as boys, that light way of looking at everything, not taking ourselves too seriously. I know that, for myself, that attitude has been my salvation. I got a lot of that from my father, just as Mac did from his father.

The other thing that we got from our dads was a sense of honesty and integrity and being there for one another. Many times I saw Tom and Joe seal a deal with nothing more than a handshake. Even though our lives have gotten a lot more complicated since those summer days on the wharf, Mac and I still hold those values very dear.

The result of Mac's phone calls is that he and I have been in more contact during the last two years than during the previous ten. We've spoken with each other many times. Last year he came across the top photo and sent one copy to me and one copy to my brother.

We had our long overdue reunion this summer at Buck's house on the Bay. The magic is still there. Thanks, Mac.

ORIGINAL PHOTOGRAPHS BY THE AUTHOR

Daddy's Little Chili Pepper

P.J. COTTON

Tatonka ♂ *SBNoM*

What joy you've brought into your Daddy's heart, little one,
I can't wait to get home to hear your voice.
The sound of those little feet running down the hall
That question you ask I'll always remember: "Daddy?!"

In your room at night, I'll always be checking
That you're sleeping soundly & covered up warmly.
My eyes mist at your crib side as I gaze at you dreaming
Wondering what you're thinking.

The walks we take together at night
with you firmly holding my finger and singing and skipping.
Asking to go to the park, yet when we get there
you want to go back home.

Watching you show off on your slide in the back yard
On your tummy, backwards, then forwards.
Holding my breath that you won't fall or tumble,
seeing the joy in your eyes as you play.

Getting you out of bed in the morning
Hearing you say "I wake up!" with such zest.
Wondering why is it that I have such trouble
leaving the covers for another day's play.

Snuggling in bed with you & Mommy
laughing and hiding and tickling each other.
Feeling you squirm & jump up on top of us,
looking for monkey who's watching us all.

Taking you to Day Care & watching you mingle
with other little ones just like you.
Thinking that it'd be fun to hang out for awhile,
never staying as long as I'd like.

So here's to you, my little chili pepper,
the spice in my life, my comfort & joy.
Here's to a wonderful future together,
whatever the Universe has in store.

METAMORPHOSIS

A Man and His Perspective

Ric Howard
Guest ♂ SCMC

There are very few things in a person's life that can dramatically change who they are. Think about it ... what have you done this lifetime that makes you so proud you can't wait to share it with the world? It's something not everyone is capable of doing, and even if they were, would choose to face. The answer is twofold: pregnancy and birthing.

As a man and father of two children, I am thankful to feel this committed, yet I am also so jealous most women have this chance of a lifetime to face themselves. What's a man to do? Where can he find the equivalent of spiritual gestation for months on end? How could he push himself past his limitations only to rejoice with tears of life and a newborn child?

The miracle of birthing can only be experienced by the conscious mother-to-be. What is my position then? At best man can hold his goddess' hands, whisper gentle words of encouragement in her ear, or breathe coaching breaths with his enlightened woman. But we want to do more, although our quest is futile. Could we as the masculine male conjure the toils of labor or muster the strength to endure hours of childbirth? Sometimes my visions place me in this holy position, and I know I can — then one day I succumb to an injury and the pain reminds me of my limited threshold, while my dream gives way to a pleasant denial. Within my pain I see no reward. Within her pain my selfishness sees a chance to sit in darshan.

I love this woman, mother of our children, and in this loving state I contemplate a vasectomy with a sadness only a man can know.

Three months later ...

... and what have I learned? A routine office procedure can wipe away the chance to see yet another smiling face. Will I ever come to that place of certainty? I've never been 100% sure of anything, and why should this be different? Am I only getting a vasectomy for the sake of my wife, and whose feelings are these anyway — mine, hers, or is society dictating once again how I should feel? Getting better at canceling appointments, that's more my style, I play the chicken well. Some days I just know deep down how easily I could go through with it — but on those days I know I don't have to.

My children are constantly in my mind. One more child, yes there's room in here for another. We contemplate a puppy instead. My mind made up today only to change tomorrow. Am I always this wishy-washy? I forget this is a major decision affecting the lives of more than just us. But this is a chance for me to show my wife how much I care about our future. Aren't the IRA's once a year enough?

I easily convince the nurse I don't want any more children, so why can't I convince myself? I'm not sure ... do I want to share my children with another, one in each of our arms is perfect. Decision making isn't my forte; procrastination is while canceling my third doctor's appointment.

Our lovemaking suffers. My memory of when my wife gave birth to our firstborn, and no one told me she'd need a long break of abstinence, is revived. This could weigh heavy on the final outcome.

Dedicated to all the children yet unborn ...

What Is A Husband?

JIM BALDOCCHI
Golden Gate Jerks ♂ SFT

One night at the San Francisco Men's Tribe Council Meeting some of the men were discussing direction. I suggested starting a Married Men's group. The idea was to get together with other married men and share our family life experiences. The meetings were open to married tribe men and their married male friends. Being married was the only requirement.

I had selfish reasons for making this happen. I got married two years ago and it made sense to me to start a support group. I wasn't sure if this project would fly, but we have just scheduled our fourth meeting.

When I talk about this group to friends and associates they often say "Well, what do you guys do at these meetings? Rag about your wives?"

Actually, what happens is we all look at marriage from different eyes. Some are newly married; some for many years. Some married when they were young; some waited until they were older. Some married and married and married, etc.

One of the men has been married to his wife for fifty years! Can you imagine? Now he's a role model. His was the generation that believed men couldn't watch their wives in childbirth because they might faint, so he wasn't allowed into the delivery room. This was totally different than what I experienced this last February when Carolyn and I had our son Daniel.

I was next to my wife in the operating room while doctors performed a caesarean section on her. Having actually watched the doctors pull my son out of my wife's stomach with the purple umbilical cord streaming out is something I will never forget. I even got to be the first person to say what the sex of our baby was. Having a child is truly a miracle. It also gave me a new respect for my wife and mothers in general.

The differences that husbands go through with each new generation is amazing. Even within the same generation the stories are unique.

So far we haven't had an agenda at these meetings, the ideas and questions just seem to flow. Who knows how long this group will keep meeting. But as long as it does there is much to share and learn.

Being married has many rewards. It is full of ups and downs both mentally and physically. And making it work is our goal.

ORIGINAL PHOTOGRAPH BY
THE AUTHOR

METAMORPHOSIS

Our Younger Brothers

DAVID MEYER
Guest ♂ EBNoM

My father left my mother as soon as he found out she was pregnant. He returned when I was three, intending to reestablish the relationship with my mother. But by that time she had married another man, so he disappeared once again.

The only other time I saw him was when I was eleven. I looked him up at his jewelry shop in Mill Valley. He gave me a ring he had made, which I still have.

I had a man — my stepfather — acting as my father, but I don't feel I had discipline in my life. I'm sure that I have experienced as much love as one could expect from a family. But as I look back I feel a lack of direction and guidance.

I had one more stepfather before I was a legal adult. Both my stepfathers loved me, I believe, as best they could and with all the depth that they thought was possible. But neither of them had ever had a child of his own. I feel there was an element missing: the feeling of responsibility that comes to a man when the child is his.

Though I received a lot of positive encouragement and was given a love of hobbies as an inheritance from the "fathers" in my life, I feel a lost loneliness in the place where Dad could have been ... a pain over all.

I implore, with all my heart, any man considering the blessings of paternal procreation to wait until your wisdom is great enough for that task (mid-20's at least). And never for a minute think that appropriate use of authority will go unrewarded.

For much of my life I have sought meaning that could have been passed down through strong bonding with male wisdom and love. That doesn't only mean Dad, but the "men's lodge" as well — the collective knowledge of the men of the community and their spiritual empathy with their younger brothers.

I had at one time in my life a man who was chosen by my mother — bless her heart — as a big brother. The time shared with that man; the memories of places traveled; adventures quickening my spirit; and his open ear and spoken heart are as much as I could have asked for from that place inside of a man.

I pray that men find time for their little brothers. Even though I have walked the road of hard knocks, teaching myself and putting blame on myself for my life in many ways, I look for opportunities to share with men younger than I. I pray that other men don't wait until the youth around them start "acting out" before reaching out to them.

Three Hawks

KIM WELCH
Chi-Pachagon Express ♂ SCMC

Three circle for my little girl.
To guide and give wisdom.

She stands and watches.
"But of course they came for me"
She says, as one who stands with
Spirit ALL the time.

Three follow my little girl.
Just to say: "Go your way."

Farewell to Russ

RICK DULA
Warthogs ♂ *EBNoM*

I went to visit Uncle Russ in a veteran's hospital one sunny day about two years ago. Going in with my mother, I knew that his life was almost over. But when I saw him that day, I realized that this would be the last time. And after sitting with him, I felt that there was something I had to tell him which had gone unsaid for many years — about a gift he had given me which had not been fully acknowledged.

Mom, bless her heart, did what her sister and mother tended to do in these types of situations: she made pleasant small talk about the nice lunch brought by the nurse, encouraged Russ to eat, and marveled at lovely day outside the sliding glass doors. Attentive, sympathetic and loyal, she did the important work of being there for a loved one. Without her influence, I would not quite have found a convenient day to visit Uncle Russ, what with a two-hour drive each way, my schedule always so full, and my denial that time was really that short.

Anyway, I give thanks to Mom's spirit which helped urge me to that room to say good-bye to my only uncle — "Unkie" when my brother Jay and I were kids.

Since Dad was a Navy man, our family of four moved around pretty often while we were growing up — Washington State, Tennessee, Northern California and back again. Summers spent in Atwater and Modesto, California, created the only continuity. Both sets of grandparents lived in Modesto and our only Aunt and Uncle lived in Atwater with cousins Ron and Sandy. This was the extent of our clan (not counting some obscure

connection to folks in Oklahoma), and these relatives represented permanence and rootedness.

The Fourth of July was usually spent in Atwater, with Russ presiding over the Bar-B-Que, and Auntie's sweet ice cream recipe churning away in the automatic freezer. Unkie was the first man I can recall doing "male bonding" and community service with the Kiwanis hot dog booth in the park downtown during the parade. I remember being somewhat puzzled by his working with these guys so that they could raise money for charities — but grownups were always doing incomprehensible stuff anyway.

As time passed, so did some of the "givens" in my life: All four of my grandparents passed away, one after another. Both cousins got married, moved away, and had kids. When I was 16 and my brother Jay was 13, he was diagnosed with bone cancer; he had his left leg amputated and went through chemotherapy and several years of nearly-annual surgeries. Uncle Russ left my Aunt; she remarried and he didn't. My parents divorced. Jay married and divorced twice.

I remember when Russ, Dad and I drove together to Grandma's funeral (she was Mother-in-Law to both of them). I was struck by these older men, how the generation they came from didn't communicate freely about their emotions. Russ' girlfriend had confided on the side that he was very saddened by the loss, and I knew that Dad must have been too. So on the way home I remarked that the pastor's tale about the loss of a mother really hit home for me, and just about made me cry. The two men didn't say anything, seemingly uncomfortable with grief. They both had hearts of gold, so when their hearts were filled with sadness which they couldn't unload, I realized that being a man carried a price.

Years later, I had done some work in which grieving was permitted and encouraged as a necessary process. Fortunately this came in handy on the occasion of that last visit to my Uncle Russ.

Although Mom's good spirit tried to bring sunshine to the hospital room that day, I knew that something was aching to get out. The more I let myself feel it, the more choked up I got, until finally I managed to say, "Mom, would you leave us alone for a minute ?" At this she (thankfully!) said, "Oh, sure, I'll just go for a little walk" and promptly left.

Russ looked at me and asked "Why did you say that?" — just as I was asking myself the same thing. Russ had been somewhat delirious up to that moment, so it seemed necessary to get to the point. In a broken-up voice I managed to get it out.

"I just wanted to tell you that when Jay was going through his operation, having his leg amputated, you sat with me and played chess, game after game for hours, and I just wanted you to know how much that meant to me."

He responded by saying, "Yes, it's good to not be alone." At that moment, Russ was completely lucid, as he hadn't been all day.

And after that he slipped back into a state of confusion and unreality. During the rest of my visit, I told him that I loved him, I had to insist that there were no cigarettes, and I helped the nurse support him in order to change the sheets.

And I was very grateful for the opportunity to thank him before it was too late.

My Secrets to a Happy Marriage

GORDON CLARK
Cool Jerks ♂ SCMC

1. Always take the scenic route.

2. Insist on time out with the boys. Don't always take it.

3. Buy a porch swing. Use it together.

4. Change the oil every 1500 miles.

5. Build a sand castle when you go to the beach.

6. Dance the dance together.

7. Don't let too many weeds grow in the garden.

8. Consider her as a special treat.

9. Don't talk too much. Listen a lot.

10. Cuddle often.

The Possibilities Still Exist

PETER DAVIS
Guest ♂ EBNoM

The possibilities still exist..........not only
For my children's dreams to work out and
For this world to be better.

The possibilities still exist......................
................that love will come home to me.

ORIGINAL PHOTOGRAPHS BY
THE AUTHOR

Billy with sister Diane
(c. 1955)

On Discovery

The world is new — for a kid ...
a stage of wonder
a place to play
to explore
to marvel
It's all right there to have!

So much to learn
to imagine
to discover
Everything is possible!

But then somehow
the innocence gets buried
deep inside
and the fears and concerns of
"being an adult" flow over
that child's-eye view
turning joy into anger
awareness into judgment
and Love into rebellion

So the task ongoing is in
re-discovering
that spirit still within
and creating life anew ...

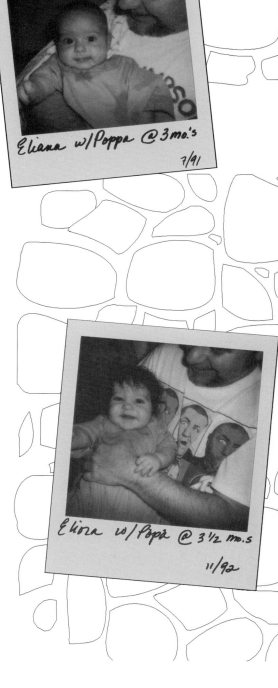

Eliana w/Poppa @ 3 mo's
7/91

Eliana w/Poppa @ 3½ mo.s
11/92

Life in Transition

BILL HORST
Jerks Out Of Hand ♂ SBNoM

METAMORPHOSIS

Gone — just like that

Bam! Like a brick up side the head

"It doesn't look good"
"We had to pronounce your daughter dead"
Bam! Gone — just like that.

Phone calls — lots of them to
    family friends teammates all
Sitting Shiva
    mourning
        Twilight Zone
    where am I?
what day is it?
Bam! Gone — just like that.

It makes no sense
It's not fair, not fair...
    Where was God, where?
"Radical discontinuity"
Ram Dass, Kubler-Ross,
    The Centre for Living with Dying
        support
    life goes on
there'll be another...

Despair Loneliness — I want to feel
    her hand squeeze my finger
        just one more time!
I explode within a thousand times...
Damn! It's not fair, not fair

Like a brick up side the head
Such emptiness swirling!
    Yes gone — just like that.
I release you
    as I cry
        Fly — just like that

Eliora Benyamina Horst
(b: 7/17/92)

Eliora — Sunshine

Sent to us by God
    on a ray of sunshine
        you lighten our hearts
As thru your Mom
    you arrived —
    dropping so delicately
        into my waiting hands
and our waiting lives

Thank you
    little sprout of life
        for choosing us as parents
In your face I see the
    promise of the sun
In your eyes
    - the future
and in your Being here
    - all possibility -
    engaged in life and
    ever pressing forward
        to discover God's world
    which is yours to
shine upon

Family —
legacy, reality
and future
promise
(Channah and
Eliora at
Eliana's grave,
4/14/93)

METAMORPHOSIS

One Man's Journey to Fatherhood

STEVE REIN
♂ EBNoM

Married at 20, divorced by 23; married at 24, divorced by 27. I was beginning to think that maybe I wasn't *quite* ready for being in the partnership of marriage during those early years: school, teaching, and construction work, mixed in with years of sex, drugs and rock'n'roll as a band manager and studio owner. Now, after marriage, my days as a bachelor appeared to be limitless.

I had my share of friends and lovers, but I just wasn't interested in risking again the commitment to a deep and lasting relationship. But then, just as I *wasn't* looking for somebody really special, Leslie entered my life — or rather her ex-boyfriend, the guy that had introduced me to her some months before, left town! Permanently!

This was a beautiful and intelligent younger woman who actually appeared to be really interested in me! So I started taking a few risks: a few dates, some all-night talks, some serious walks. I really let her see inside me. To my initial surprise, I didn't fall apart. The relationship didn't end; it grew, I grew, we grew together. I moved into Leslie's apartment and walked away from a lot of the old extra baggage I had been carrying around inside me for years.

Suddenly, it was more than three years later, and I had made it past that magic number which seemed to doom my previous "serious" relationships. The sun still shone in the daytime and the stars at night, but my life was very different. In fact, all things considered, life was great!

But soon, the inevitable "M" word started to be alluded to. It seemed to be part of the natural progression, but I was so happy to have finally found a great partner, companion, friend and lover

that I was scared to think what might happen to our relationship if we got married. Sure, I had done it before, even though it seemed to have happened *to* me, in another life, in another world.

Leslie, still a younger woman by my standards, had never been married, and the hints were difficult to miss. I successfully dodged the issue for a while, but finally my confidence in our relationship overwhelmed my fear of change, and I decided to take another risk.

Well, we've been married almost five years now, and they've been the best years of our lives. Life couldn't have been better. We've worked at good jobs; traveled to Canada, Mexico, Europe, and Asia; we skied and rafted and bicycled and danced; we explored, and moved, and loved, and I was sure that this was a great combination for the rest of our lives.

Then that other "M" word, "motherhood," started popping up at intimate moments. With twenty-something years of adult life under my belt, a great life with my wife and partner going on day after day, and no great paternal urge having pushed its way to the fore, I thought it wasn't a big deal if I lived the rest of my life without being called "Daddy." But Leslie wanted to "at least talk about the possibility."

Well, now the "possibility" has *almost* become history; our little girl is due in mid-April and I'm sure that this will be the most exciting part of our lives together. Sure, I was hesitant about another change in our lives, scared that we might lose our wonderful relationship, that things had just been great for almost ten years and that now they would be different forever.

In fact, **things will be different forever**! Every parent we know has told us so. But I think I've finally learned that only by taking the risks and having your life change do you really grow and find that there are always new ways to have a rich and wonderful life.

I'm more excited now, even if still a little apprehensive, about this enormous imminent change in our lives than I have been about any of the previous changes in my life. With more life experiences under my belt and a wider circle of support systems, I'm also better equipped with a healthier perspective than ever before, and feel that we have a whole new and wonderfully challenging life ahead.

I'm going to be a great daddy!

(Editor's note: Elena Perry Rein was born April 23, 1993. Mother, daughter and daddy are all doing fine.)

What Nicholas Told Me

CHRISTOPHER DINGMAN
Sperm Whales ♂ SCMC

I put my sister's child to bed.
Two years ago
he hadn't entered our world.
Now as I lay him down
on the pillow
he looks up at me and says
"I'm going to dream of bats...
And balls, and suitcases, and cameras."

I look at his fine blond hair
where it falls
on his small forehead.

The Last Backpack Ride

BRUCE KING
Bushwackers ♂ EBNoM

When my son Casey was two months old I started carrying him in a powder-blue Gerry backpack.

I often had to forcibly smooth my stride to keep his little head from bouncing against the frame as I carried him up and down the Oakland hills.

At first, Casey would fall asleep soon after we set out. As the months passed, his baby sounds gave way to individual words, then phrases.

I grew used to Casey's bouncing on my back with increasing vigor as he became larger and stronger. We began talking to each other, and he would laugh as I repeated his mispronounced words back to him.

Our Gerry pack became stained, torn, and battered with constant use, and was finally replaced with a larger, tougher model which Casey is rapidly outgrowing.

My Casey is no longer a baby or toddler. He is now a little boy of five years and well past forty pounds, and our backpack rides are becoming less frequent.

Lately, when I take Casey for a "hike," I wonder if this will be the last time that I feel my beloved son on my back. Will I hear for the last time, "Daddy, Daddy..." from behind, or feel for the last time a little hand playfully pulling my hair or covering my eyes? Will this be our last backpack ride? Where did my little boy go?

One day it will be time to throw away that second battered backpack. But the memories of those backpack rides will stay with me forever.

Transformation by Death

LARS RIDER

NADS ♂ SBNoM

The following four poems expose my pain at losing my father, mother and brother. My father died at the end of my thirteenth summer, two days before school started. He died from diabetic complications. My mother died when I was 21, succumbing to cancer. My brother took his own life, seven years ago.

With each poem, I tried to convey my feeling and the transformation that occurred with each death. I have found that dealing with these deaths transformed me more than any other events. My father's death made me a loner, an outsider. My mother's death gave me love, tolerance and understanding for others. And, my brother's death gave me ruthless compassion for those around me and a very dogged need to be persistent (y'all may have noticed this already).

Finally, "A Speck of Foam" illustrates my feelings with all of these events and at the realization of my (and every one's) separation from each other. Thank you, to my team and the men of NOM, for making this separation more bearable.

ɤ

THE BOY WHO WOULDN'T CRY AT HIS OWN FATHER'S FUNERAL

He asked for wood and they gave him gray steel.
He asked for full honors and they gave him "the standard."
He asked for a bugler and they gave him a recording of "Taps."

But, no one would listen to the boy who wouldn't cry at his own father's funeral.

He asked for their understanding, and they gave him their fear.
He asked for their love and they gave him their hate.
He asked for their respect and they gave him their scorn.

But, no one would listen to the boy who wouldn't cry at his own father's funeral.

I told them, "A wood coffin" and they patted my head.
I told them, "Gerard plays the bugle" and they patted my head.
I told them, "A rifle team" and they patted my head.

No one would listen to the boy who wouldn't cry at his own father's funeral.

So I cried inside at the cold steel casket.
So I cried inside at the lone rifle firing.
But I cried out loud at the bad recording of "Taps" he sang nightly.

Because, no one would listen to the boy who wouldn't cry at his own father's funeral.

METAMORPHOSIS

WHAT NEEDS TO BE HEARD

I look into your hazel-gray eyes,
deep pools of wisdom and calm,
I see the discomfort, the love and the waiting,
to go to your maker, your stay with us ended.

You accept this unkind fate,
of pain, hopes cut short,
with a grace that's amazing,
as it gives me your warmth.

So fragile you seem,
your last breaths coming,
that I think I could lift you,
with no trouble at all.

As your eyes roll back,
I think of the voice that had called me;
"It's Time, It's Time," and I ended my work
to be with you as you started
your journey.

Your last breath taken,
your body is rattling.
Holding on, holding on
while you're already gone.

I weep very softly,
caressing your hand.
I've never known someone
who cares like you can.

After a time,
I go get your daughter, your sister,
to see that you're gone and
to say their good-byes.

Then I go call your sons,
my brothers, to come
and see their kind mother
in gentle repose

Upon my return,
My sister has a message.
You sat up and told her
to tell me "Be Good."

Though you were gone,
you showed me the way,
that love can lift you
or return you to say

What needs most to be heard.

A SPECK OF FOAM

A speck of foam in a sprayed ceiling
Asking nothing, receiving less
Always seen, yet overlooked
In a crowd, yet always alone.
God, what solitude,
Rider.

METAMORPHOSIS

ORIGINAL PHOTOGRAPHS BY
THE AUTHOR

FOR MY OWN SAKE

Down dark paths we traveled together.
Sick with the loss that life can hand you
I was the one who got frightened into ending my farce.
You were the one who found the Abyss too friendly.
Our paths diverged, with me getting stronger, and when
I tried to lift you up, you proved much too heavy.
Then the unthinkable, the horror, it happened.
I let up on you and you finally sank under.
I blamed myself, I wasted myself, I hurt myself
to make up for being a bad brother.
But the facts remain, I threw you a rope.
Now I've learned not to give up, and
I'll go way too far, just not to sell out
Some find fault in this, and occasionally scorn.
But it's my way of honoring you,
of keeping you alive.
You killed yourself once, and I can't do it again.
For my own sake.

Home

KURT ANDERSON
Bucks ♂ SFT

The women are gone
out of the house
off to work.

The home is empty
and full of dust.

A coldness lingers
through our hearts.

The war of sexes is engaged
No one gives in.

Just the corporations profit ...
from our family
our values
our souls ...

Finding A Lost Journey

JEFF PETRILL
Landsharks ♂ SCMC

My dad announced he was taking us to an amusement park. At first we couldn't believe we were really going. The news came as a sudden surprise, and all of us kids quickly cooperated to make this dream ride come true.

We packed and got everything prepared. It would be a long journey, but worth it. The trip was from Flint, Michigan to Sandusky, Ohio and we were going to Cedar Point, which was like a Disneyland for that part of the country. On the way, we sang songs, ate sandwiches, did our best to get along because somehow we earned this unbelievable treat. Or maybe my dad planned it long ago. Setting off early, we endured the many miles, and then we found it.

The day was full of rides, sights and sounds that I thought I would never forget. But it's been so long, I only recall a few images, mostly a boat ride with fake animals that popped out of the water and sprayed us. What I vividly remember is the ride home. We got lost.

After a glorious day in heaven, we rode off into a night of hell. Looking back, it wasn't nearly so bad, but I was too young to understand my own expectations. It started when my dad said, "Is this the way? Wait now. I think it's ..." And nobody knew what to say or really paid any attention, even my mom, until she finally said, "Jack, pull into a gas station and ask." But he insisted, "I know where the hell I'm at!" About five minutes later he said, "What exit is this? Wait a minute. Where's this road go to? Oh, I think I know." And later saying, "What the hell ... ?" Then we joined in with our mom, "Go to a gas station and ask somebody." But he didn't do it. He thought he could figure it out.

As we got more lost, we got more tired. Brothers and sisters can get along for a great day at a great amusement park, or at least show remarkable improvement on how to be together, but as one kid slouched into another, the whining began, and then fights, and then my dad shouting, "What the hell's going on? Do I have to pull over?" And somebody stupid enough, such as myself, sarcastically replied, "Yeah, to a gas station. You don't know where you're going." My dad got pissed. He kept driving into a black hole, and we kids had nothing better to do than fight and blame each other for making a long trip longer. Nobody could sleep. A car can only go so far, and becoming a motel wasn't part of its gears. I can't recall any ordeal as long and tiresome, a trip with no end, just a rolling miserable series of complaints. Everybody knew the next day would be shot. We had to get up early for something on Monday, probably school, and it scared us to think we would walk in as zombies. "Our dad got us lost last night!" would be the story to excuse our wretched selves. That is, if we ever got home ...

My mom grew weary of the circle leading nowhere. "Honey, go to a gas station and ask for help. Geez!" By this time, my dad was too angry to say anything. So she repeated it, and then the earthquake hit. "What gas station? Do you see a gas station? Show me one!" We had rode off to another planet, dark with no lights, past everyone's bedtime, and somebody's elbow kept poking into me so I poked back and somebody hollered and my dad yelled, and when he calmed down he softly said, "Hmm. I thought this road would lead to a different one. There's probably a gas station somewhere ahead. We'll get there. How did you guys like that boat ride?" He was answered by a baby crying, woken up suddenly, probably discovering in baby language that we were lost.

I don't remember when we got home. The house was just suddenly there. I opened my eyes, after the approximately fifteen minute sleep I got, and muttered, "Finally." But the good news wasn't so good, since now I had to help carry in the little ones who miraculously fell asleep and couldn't wake up. And I had to help carry in all the other stuff too. I was my dad's helper. But I didn't like it that way. He should've pulled into a gas station way back when! And he noticed I wasn't the most cooperative helper in the world. And he told me so. When I was released for a good night's rest, it was almost time to get up. A new day awaited. I was angry and felt ripped off. The great time at the amusement park was forgotten. My focus was not having all the rides the way I wanted them. Too many turns and twists that got me lost in confusion. I felt bitter.

Of course, nowadays, when someone in the family says, "Remember when dad got us lost coming back from Cedar Point?", we all get a giant laugh. And my dad will make one of his funny faces and shrug.

I know he really believed he knew the way home. I know at times he got scared because we were so far off the path. But he got us there. Somewhere in my brain I would chalk up his mistakes. Anger and confusion keep score of the strangest things. But humor and insight know better games. I was on a ride I had to be on. He's the one who took us, made it all possible, did his best and steered a straight line, no matter how tired and how far. I rode through parts I found difficult. That's the way it goes. And as I drive through life, taking others with me, responsible for my actions, I can hear some complaints or voices who expect something else. I think back to a big ride coming from the Disneyland of the Mid-West. Hang in there, as my dad always said. I keep driving, and listening, and going forward.

The long road leads somewhere. It really does.

Family

DAVID W. NEHRKORN
Grunt 'n' Pokers ♂ *SBNoM*

What does the word *Family* mean to you? Webster's says that it is derived from the Latin word *familia* which means household and includes servants as well as the kin of the householder. If you looked in the dictionary, you would find: "the basic unit in society having as its nucleus two or more adults living together and cooperating in the care and rearing of their own or adopted children." Unfortunately, the last decades of the twentieth century have seen a dramatic rise in the divorce rate (especially in California). Perhaps this is because divorce is now socially acceptable. Parents used to remain together "for the children," even though they despised one another. Does this situation benefit anyone involved? I don't think so. We need to forge a new definition of the *family*.

To begin this process, let's ask ourselves who are the most important (and vulnerable) individuals in a dissolving marriage? The children, that's who. They often think that everything is their fault. They frequently get neglected, cut adrift, and are left to fend for themselves. A new family unit needs to be created from the ashes of the old union. In this unit, both parents (and new spouses) are equally involved and committed to the rearing of all children, including those yet to be born. I am not talking about some new age, communal relationship, but an "extended" family. What I envision is very achievable in today's society. It takes commitment, compassion and understanding — perhaps more than you think you are capable of. But it is well worth the pain of transition.

This goal must have its roots in the divorce process itself. Divorce itself is not necessarily evil; but **you** must make the outcome positive. After a separation, life is a battleground. There is blame,

METAMORPHOSIS

and there is a lot of hate that reaches its peak during the painful process of dividing accumulated resources. No one is at fault — the relationship was at fault, and no blame should be cast. Arrive at a settlement through mutual agreement! The courts have tables for determining alimony and child support. Don't involve lawyers; they only escalate the hate spiral. Once a settlement has been reached, you can begin the healing process. Make this happen as soon as possible; any delay gives bitterness a chance to develop. Then, every contact is a confrontation. Once established, this pattern is very difficult to break.

Offer to pay alimony, now called "spousal support," in one lump sum. This arrangement has advantages for both you and your spouse:

1) By offering a single payment, you can probably get your wife to accept slightly less than she would get with multiple payments. After all, $15,000 in cash is a lot more attractive than $500 per month. It can be used as a down payment on a house, for example.

2) One of the conditions of alimony payments, is that they stop when she re-marries. Now, she can marry at any time. And frankly, like it or not, a family with two parents is a much better environment for your children.

3) And finally, probably the most important item, a monthly alimony payment is a constant reminder of the old wounds. It prevents constructive healing. On the other hand, think of child support as a benefit payment for your child, his or her "paycheck."

Don't be inflexible with child support. There are some duties that won't come under the umbrella of a routine payment. For example, if you think your child deserves a private rather than a public education, be prepared to split the expense. Your child's education is an investment in his future.

Pick up the pieces, and leave the past behind you. Don't fight for custody; a custody battle is difficult, expensive, and ruinous for the kids. Think of what is best for your children. Are you fighting for custody because you feel that your ex-wife is an unfit mother, or are the children some prize in a game? I recommend that you choose joint physical custody with primary residence with your spouse. You will be part of any important decision regarding your children, and the court is taken out of the visitation loop. Visitation will be based on convenience (yours and hers), not an arbitrary schedule. Your spouse must believe that your child **needs** to see his father regularly.

With the separation details taken care of, it is time to build a new relationship. What relationship, you ask? The one between you and your ex-wife's new husband. He will be your child's new "father," and with luck, he will be a dad. Encourage him to think of your child as his, and to discipline and praise accordingly. I know that this will be a difficult thing to do at this time, but remember, he will be the parent-in-residence most of the time. It is up to you to forge a cooperative relationship. At some time, children will come to this new marriage, and the principles I have outlined above should apply here as well. Involve your parents in this, too. It's a natural; grandparents love kids. You can truly make this a synergistic relationship. I tell my son that he is a lucky boy because he has two dads that love him, and most kids have only one. Think of the advantages of this extended family. For example, my son's new dad takes him skiing, and I take him bike riding. I teach **both** of them about Macintosh computing.

This is definitely a new definition of the nuclear family. Some people have said to me, "Oh, how California." I say, "Innovative solutions for 90's relationships."

Injury and Childbirth

TIM BATESON
Warthogs ♂ *EBNoM*

This is a story of my journey through injury, through conceiving a son, to the realization that I had to do something about my, as I would call it, predicament. I. Me. None other. First, foremost, most importantly, most obviously, to leave the relationship I had been mired in for so long, a seemingly endless seven years (I know it fits the cliché).

What I *was* doing at the time was riding my bike — obsessively. As a friend and mentor pointed out to me at the time, it was my way of escaping. This seemed to fit. I could ride for hours every day. Finally I rode to the point of injuring myself from overtraining. I developed bursitis in my hips. The doctor said to cease all bike riding immediately or risk being permanently injured.

This I did and huge changes in my life were triggered. Not riding, I found, left me with an incredible amount of energy that I could use for doing many things. I did not realize how much I was putting into my bicycles. I started doing things I had long been missing. Or avoiding perhaps. I rediscovered the joy, and challenge, of meeting people, spending time with them. I rediscovered that relationships could be a source of great joy, not just stress and conflict.

20 Tim Things
Calligrapher · Warthog
Letterwriter · Listener
Son lover · Adventurer
Womanchaser · active
Bikerider · Openhearted
Handyman · Political
Handsome · Courageous
Skier · Parent writer
Divorced · World tra
Outdoorlover ··· veler
Recycler · Graphicartist
Nude bather · Commited

Suffering in my desperation to change my life, but still not really able to consider leaving the relationship, I decided it was time to have a child. Not a good time, certainly not the best time, but something I wouldn't put off any longer. I felt that I had been merely putting it off, partly because of my partner's almost daily requests to have a child.

At the time, thinking I wanted a kid was the closest I could get to knowing how I felt, what my feelings were, or that my feelings mattered at all.

The birth of Christopher (now much loved) was, however, a strong enough kick in the ass to bring me to my senses, to my desires, to my feelings, at least close enough to them to believe what had long been nagging at me from far back, deep down — what I really *did* want was, out of that relationship.

Through all the pain, self-blaming and shaming, and all the difficulty of separation, I have always felt that my life was finally starting, and moving in the right direction. I have done things, accomplished things, I *am* accomplishing many things that I had only dreamed about for so long, had imagined were only for other people and not for me.

Life Goes On

HARVEY ROSEN
Bushwackers ♂ *EBNoM*

If you have yet to experience it, you will. What I'm referring to is the fear and pain we experience as we watch our parents age and die. They are more than just a reminder of our own mortality; these experiences touch our core with the fear of loss and the recognition of the trials that we must face alone, without them.

Eleven years ago my mom got real sick. "Prove to me she has cancer," I remember yelling at the doctor, totally in denial. I also remember being angry with my father, who was there as we watched my mother's failing health. He seemed so calm, so composed, as my heart was being mauled by these events. I remember confronting him with my anger over his composure. As a survivor of the Holocaust, his response was simply, "Son, this is not the worst I've seen in my life."

.I knew Dad dearly loved Mom but I also knew of the shit the woman would dish out. I'll never forget the day before she died, my dad and I visited her in the hospital. My mom loved ice cream so my dad brought some for her. When he gave it to her she laid into him. "You're an idiot!" she screamed, because he had gotten the more expensive kind and should have used a coupon to save money. They had a unique relationship. I asked him how he could take her for so many years. He just looked at me and said, "In one ear and out the other."

I traveled to New York five times during those months. As my mother's health failed I began to get more concerned about my dad. I'd heard that sometimes when people spend their lives together the surviving partner loses his/her will to live after the loss of the other. "How will Dad deal with this?" I thought. He was a man who had never been to a grocery store or washed his own clothes. It wasn't that he was lazy or stubborn, he just didn't need to; my mom handled all that.

It wasn't until six months after her passing that I allowed myself to feel the pain. It hit me like a ton of bricks as I drove down Highway 101. I pulled off on the shoulder of the road and just cried.

My relationship with my dad was transformed after Mom's passing. The relationship had to change; he was now a different person. He had always been like a shadow in my mom's presence. Now he was getting out, going to synagogue, socializing with his peers. It touched my heart because I realized that for the first time in my life he seemed to really be enjoying life.

When I think about her, I remember mostly that my mom and I had some tough times, but I really miss her. I'll always regret her passing and remember her with love in my heart. And I'll always cherish the relationship with my dad that was only possible with her passing. Thank you, Mom. Thank you, Dad.

A Painful Lesson

JOHN PETERSON
Quiet Riot ♂ *SBNoM*

Dear Men:

Recently and abruptly my wife of thirteen years said she no longer wanted to be married to me. This is the same woman who until recently said "quitting is not an option" and that she was committed to me and our two-year-old son for life.

I felt blindsided and betrayed. I have never experienced pain like this, not only due to the tremendous loss, but to my lack of understanding. How could this have happened? The relationship felt so secure — just days before she left, she treated me like the hero I thought I was to her. There was no reason, and no clues — or so I felt.

Then one day I received a letter from her father. Among other things it contained the following paragraphs:

> "First off, in my own case I find that as I grow older I hear myself sounding more like my Dad! When I do it, if I become aware of it, I am horrified because here I am, doing and saying things which I have sworn never to do or be. I think that this is a common enough occurrence in adult life of the male, that you are probably finding yourself doing the same thing with your own Father. Watch out! This can lead to trouble."

> "In dealing with women, I have found that there can never be enough romance — not necessarily sex — but the little things that reassure the lady that she is still the beautiful person that she wants to be in your eyes. Try on a daily basis to notice in a positive way her clothes, her physical beauty, the things she does around the house — decoration, food, art — 'spirit lifters.' "

Whether he had a window into my soul, I don't know, but the impact of his words reverberated through me. The "things which I have sworn never to do or be" brought back memories of my father and how he had totally dominated my mother, forced her to be submissive and insecure. I had resolved that I would never do that in my relationship. Yet, as a man ruled by Clint, was I doing the same thing? Was docility my wife's natural being, or the pose she adopted to deal with my domination? Were her efforts to show me I was her hero mirrored by mine reminding her she was my queen?

As men, by nature or even teaching, we tend to leave relationship to the women, as though it has nothing to do with us. We think it will work out by itself, or that the women will fix it when it needs fixing. So instead of doing preventative maintenance on the relationship, we are constantly doing damage control. Sometimes, the damage becomes irreparable — and there we're left, hurt and bewildered.

Perhaps the clues were there all along, but I was too caught up in my upbringing and macho act to see them. Men, relationship is precious and fragile. Get off autopilot and be present, before you miss the last clue. If my pain can help just one man, my experience will at least not have been in vain.

Love,
Peterson

METAMORPHOSIS

EXCURSIONS

METAMORPHOSIS

Marked for Life

Henry Block
Bushwackers ♂ EBNoM

It's a mark — a physical mark on my upper left thigh. It's made with a stainless steel guitar string needle, using imported German ink for the stain. It's hard to describe this figure, which I would call a tattoo; needle artists would call it a tribal tattoo. It's even more difficult to explain the meaning this mark has for me. It represents more a state of mind than a geographical place.

Aboriginally, men had these marks put on their legs in sets, ankle to hip, inside and out. They were applied using bird-bone needles, and lamp black as the ink. These men inhabited the island of Pulap, about 700 miles south of Guam, seven degrees off the equator. I lived with these people for two years, and I have been marked ever since.

The mark itself represents a set of three dolphins. I've heard three stories as to why the Pulapese honor these mammals. Dolphins are often seen traveling with the traditional ocean-going canoes; cavorting and leaping with abandon in front of these vessels, they seem to be escorting or leading the travelers. Secondly, we never see sharks when dolphins are in the area; are dolphins protecting us? Thirdly, there are tales of drowning men being held up and saved by these mammals.

The marks I carry represent a combination of the island, its people, myself, and the time in my life I resided there. I was a young 21-year-old college student looking to put off the inevitable draft notice that my friends were receiving.

The Pulapese welcomed me with open arms, housed me, fed me, trusted me, loved me, and treated me as a respected elder. As the first Western person to live among them, I did my best as a schoolteacher, as a community development worker, and as a man. Looking back, I am amused at my feeble contributions; but when I dwell on what they gave me, I am touched in a very special place.

The people of Pulap accepted me as a brother, son, uncle, cousin, clan member and friend. I had a "mother" and "father" to look over me and teach me what a man needs to know to be a man. Occasionally they were blunt in their instructions, but mostly they did it in such a way as not to be intrusive. The boys who were my students during school hours were my teachers in the ocean, in the jungle, in the trees and on the canoes. I guess I was like a crazy uncle, not quite all there, who needed to be followed around so he wouldn't get into too much trouble.

The tattoo is not the only mark I carry from my Peace Corps experience. Spiritually, I was impressed by the islanders' harmony with their world, especially the sea. To this day when I taste sea water I remember them.

In the moral realm, my beliefs regarding war, especially Viet Nam, were molded by my two years on the island. After my service, I returned to college and earned a graduate degree in anthropology, so much was my need to further understand the islanders and their way of life.

Politically, I came to understand that democracy is not the only form of government that works. A new understanding, first hand, of terms such as *matriarchal*, *matrilineal*, *clan*, *wife's uncle*, and *chief* came to me.

Physically, beside the tattoo and various scars from infections, I lost a tooth, and I felt the pangs of hunger when food was scarce for all on the island.

The jury is still out on how I view religion since my stay on Pulap. Missionaries converting the noble savage can cause mayhem culturally, but it seems to bring them joy now. Let it be said that I observed interesting religious practices, both Christian and traditional, that still influence my relationship with God.

I like the way my marks look, and I like the way I feel when I think about them. Last year I took a trip west, a pilgrimage so to speak. It had been 25 years and much had changed in that one generation. When my feet struck the sand of the beach, multitudes of feelings returned. It was as if I had returned from a short visit. My friends on Pulap received me right back into the community and welcomed my wife regally. I was cared for and loved in a special way by very special people.

I am happy and honored to be marked as a man from Pulap. I carry my dolphins with pride.

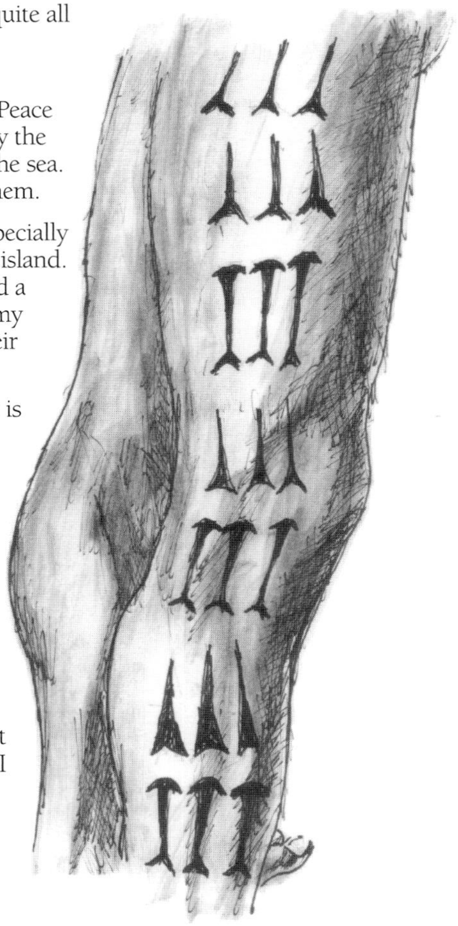

Original pen & ink drawing by
Lee Bragg
Guest ♂ EBNoM

The Saudi Man in His World

JOHN RENISH
AaaRrrr! ♂ *SBNoM*

The Saudi man is one of the most commonly misrepresented and almost totally misunderstood men in the world. The same is largely true of Saudi women, but that matter will have to wait for somebody else to take up. Saudi men are, by and large, upright, gracious, hospitable, loving, and gentle.

Wait a minute! you say. Are we talking about the same lust- and violence-crazed oil sheiks we have all come to know and loathe? Well, yes. And no. After having lived among these people for nearly two years now, I have yet to meet an oil sheik. Come to think of it, I have yet to meet a sheik, mostly because I don't travel in the heavy-duty religious circles.

What? *Religious* circles? Well, yes. *Sheik* is a term usually applied to a man who is recognized as pious and learned in religious matters. Sometimes Arabs use it to apply to a temporal ruler (the King is, rarely, called a Sheik of Sheiks), but usually they mean somebody who is a model of righteousness. In our culture, a sheik would be perhaps a Billy Graham or a Jimmy Carter, but definitely not a Jim Bakker or a Richard Nixon.

So, who are all these bearded lunatics in sunglasses? Mostly, they're about as lunatic as Ward Cleaver; living in a desert climate, they find sunglasses just as convenient and comfortable as we do. In Riyadh, though, a Westerner is much more likely to be wearing shades.

Our image comes primarily from our popular media. For example, if a Western political cartoonist wants to represent Everyarab, he draws somebody in Gulf Arab headgear, bearded, big-nosed, sunglassed, and usually with an evil leer or sneer. The same Western cartoonist would lose his job in a hot New York second if he drew Everyjew as yarmulkahed, bearded, and big-nosed, hold the sunglasses, but keep the evil sneer.

So, bearded they are, but lunatics (with some notable exceptions) they are not. In fact, the question of beards came up in a conversation I had with some Westerners the other night. One asked, whether rhetorically, wimpily, or hostilely, I'm not sure, why it is that members of orthodox or conservative religious movements tend to be bearded. I responded that most likely it's because they are from cultures in which being a man is not a Bad Thing. In our own culture, for example, beards have been fashionable for a much greater time than they have been out of fashion. Perhaps it is not coincidental that beards went out of style as the women's political movement (first Prohibition, then suffrage) came into being in the last century.

Beards here are less common among younger men than among older, except for those who want to be perceived as more religious. These young men tend to have a legalistic and judgmental approach to life, contrary to the tolerance Muslims are supposed to hold toward others. Unfortunately, their numbers are increasing rapidly as they graduate from the religious universities with no skills fitting them to gainful employment. I don't travel in their circle, either, although in social meetings they can be quite pleasant.

✑

Despite the generally negative view of the Arabs our culture conveys, their reputation for hospitality is high in the West. Not surprisingly, their graciousness and hospitality are even greater than the reputation they enjoy. When Richard III (The Lionhearted) lay ill with fever in Damascus during the Crusades, Salah Al-Din (Saladin) sent runners to the mountains to bring snow to cool his enemy's brow. These are the men who invented chivalry, of which the Europeans had only a pale imitation.

METAMORPHOSIS

For a more current example, our espresso maker blew an O-ring, so Laurel and I took it to the distributor for repairs. While I was explaining the problem to one of the foreign hirelings, most likely a Palestinian, a Saudi came from another office and invited us to join him for tea. He asked if we liked the machine, determined that we had bought it in the States, and then carried on a long and pleasant conversation with us about business, life, our respective families, and personal relationship and its primacy (especially over business). Turns out he is the owner of the company, and he is the sole distributor not only for Krupps products but for Bayer products (chiefly agricultural chemicals) and for several popular brands of cologne, in a country where all the men smell better than any woman I've ever met. He is also a Quraish, a member of the same tribe as the Prophet Mohammed, and the most prestigious tribe in a country where tribal affiliation is crucial. When we returned to pick up the machine a few days later, there was no charge.

Two symbols of Saudi Arabia, more typically Saudi than even the sword and the date palm, are the coffee pot and the incense burner. It seems every other building or public fountain features a coffee pot or an incense burner. You cannot enter a Saudi's house without being offered coffee or tea and some dates or other sweets. Similarly, you cannot refuse to drink at least one cup of coffee (barely more than a thimbleful), nor can you drink more than three cups of coffee. The tradition goes back so far that it is supported by a manly little folk tale. It seems this Bedawi (plural is Bedu; Bedouin is a French corruption of the Arabic) had a particularly beautiful wife. When a man came to his tent, he offered his guest first one cup of coffee, then a second, then a third, and yet a fourth. The guest's taking the fourth cup confirmed the host's suspicion that it was not coffee the guest had in mind; so the host struck his head off with a sword.

Hospitality is deeply ingrained both by tradition and religion. Even a blood enemy, one you have the right to kill, can demand hospitality and protection, in the desert, for three days; and he has the right to three days without molestation after leaving your tent. During his status as a guest, you have the obligation to protect his life, even with your own, regardless of whether he is a close relative, dear friend, wanderer, or sworn enemy.

Speaking of the Bedu, they occupy a strange place in the Saudi popular imagination, something like cowboys and Indians do in our own. They are seen as primitive, poor, uneducated, and a little disreputable. Yet at the same time, they are known for speaking the purest Arabic, having the most subtle culture and personal interrelationships, and being uniquely free in a progressively stifling world beset by foreign ideas and material goods. Regardless of their wealth, the Bedu honor generosity above all. About 40 years ago, the most revered man in the Empty Quarter, a vast and inhospitable place even by Saudi standards, was also the poorest of his tribe. His claim to fame? Every time guests came to his tent, he would slaughter a camel. He continued this practice until he owned nothing but the clothes on his back and six goats.

Saudis dine differently from us: the coffee, tea, and conversation come before the meal and go on for hours. After eating, the host lights incense, with which all his guests perfume their hands, faces, and headgear. They then leave because "Badr a'ood, mafi ga'ood." (After the incense, there is no sitting.) The hostess treats her female guests similarly. It is unusual for the man and women of a household to entertain guests at the same time; in any case, the parties are always strictly segregated by sex.

This business of sexual segregation is very different from what we think it is. Westerners tend to see it as backward, insulting to women, sexist, and so on. Saudis, however, see it as a necessary protection for women from the lust of those men unable to control it. While each man might feel he has some handle on his own lust, he's not sure about other men. At the same time, he would just as soon not have his own control tested. So, the Saudi man prefers women in his charge (wife, younger sister, daughter, maybe mother) to wear obscuring clothing, even a veil, to protect everyone: the women from unwanted advances, and the men from the temptation to sin.

Despite the general segregation of the sexes in public life and during entertainment, within the family, men and women are usually very close. This is difficult for a Westerner, particularly, to understand because Saudi marriages are always arranged. When a man decides it is time to marry, his family finds another family with a daughter who will be an acceptable bride. Acceptability is a matter primarily of the reputations of the individuals who would marry and the reputations of their respective families. Even mild criticism, such as, "Some people would consider this family is not suitable for you" would end the negotiations. It is not entirely a blind selection, however. If the principals have particular desires, for example, if the man says he will not accept a bowlegged wife, his mother, aunt, or sister can go look at the legs of the intended bride. Similarly, both the man and the woman are entitled to see the faces of their intended spouses. And finally, the wife has the right to a dowry payment from the husband, consistent with her social standing. The dowry is a legal debt that may be paid before or after marriage, at the bride's discretion. If the husband should die without paying it, the dowry is a legal encumbrance on his estate, in addition to the legally required portion she receives as a widow.

Saudis woo their spouses after the wedding, and they do it with a vengeance. Along with chivalry, the Arabs also introduced rhyming poetry and the notion of romantic love to the West. A classic Arabic poem written about 1500 years ago extols the beloved's beautiful eyes and plump ankles. My lady's opinion of the Arabs went up several notches when she read that. Only about two to three percent of Saudis of the current young generation take a second

wife, and taking a third or fourth wife, while legal, is virtually nonexistent today. While divorce is possible, it is frowned upon; and if the husband initiates it, he must still pay the dowry. The marriage is seen as a relationship between the families involved. Also, children over the age of seven belong to their fathers, so everybody has a vested interest in maintaining the marriage on equable terms. One Saudi told me his compatriots are very successful with American women simply because by comparison American men treat women like a piece of furniture.

Most men you see alone with their children in the States have that haunted look only weekend daddies know. By contrast, if you go to one of the many small amusement parks in Riyadh, you are much more likely to see men than women with their children, indulgence and love written all over the fathers' faces.

The gentleness of Saudis may be the greatest surprise to Westerners. Despite the relatively recent experiences of nation building and urbanization that took a herding and raiding economy into the late twentieth century in less than 20 years, these men are remarkably reluctant to engage in even verbal arguments. Egyptians are in-your-face types with practically no provocation, but Saudis get hot, seemingly, only following one of the distressingly frequent traffic accidents. Even then, a crowd of perfect strangers (men only, of course) will converge on the scene to keep the drivers apart.

Any American who comes here comes with a load of mental baggage. Fortunately, I'm a very light traveler in every sense, and I have learned to admire, trust, and even like these men, though I still find some things about them quite baffling. God willing, the next two years will bring as many revelations.

Rwanda Journal

ROGER BEHNKEN
Wailers ♂ *EBNOM*

July 6, 1989: We are assigned gorilla group 13. We begin our trek with six people led by two guides — one carrying a rifle. We walk for an hour and a half along a trail. Our guides pick up a scent and lead us into dense jungle. We are bitten by fire ants. I slip and grab onto some stinging nettles; my palm feels like it's on fire. I curse my girlfriend Ellen for coming up with this idea of chasing gorillas.

After three hours our guide stops and tells us we are among group 13. I don't believe him until he pulls back some foliage, revealing a 600-pound silverback sitting there eating berries. His family of three young males and two females with their babies are in nearby trees; they are also having lunch.

We have only a short time to shoot pictures of our closes relatives in the animal kingdom. The silverback approaches me and nudges me with his shoulder into the brush. I am shaken. The guides say it is time to leave after only 20 minutes. We silently retreat.

ORIGINAL PHOTOGRAPH BY
THE AUTHOR

Two Lessons

PHILIPPE LAVIE
Happy Men ♂ *SBNOM*

Once upon a time, when time was not as important as it is today, actually when time really didn't matter much but for the animals, I encountered Jacob, a wise old man, hairy and tall, proud and majestic. I stopped for a while. At first, he would not talk much. I realized that I had to be reviewed, probed and accepted as a real person. Whether I was or was not did not matter, I had to go through the process.

And one day, the day of my departure, Jacob pulled me aside as he wanted to share with me an old tale from his home far into Eastern Africa. I really had run out of time and had to go on. Nevertheless, I took the time and listened to his story.

I was surprised as it came quickly and directly. Here is what he said: "Every morning in Africa a gazelle wakes up realizing he has to run faster than the fastest lion or he will be killed. Every morning, a lion wakes up realizing that he has to run faster than the slowest gazelle or he will starve." He went on to say, "It doesn't matter whether you are a lion or a gazelle, you better wake up running."

We both laughed our hearts out, and from that moment on I always remember Jacob's story when I am running or when I do not take the time to look at the sky and the birds, or breathe in the fresh air of our world. It's at that magic moment when my past comes to my present to benefit my future, and I remember that time really does not exist.

A most wonderful encounter happened just a few days later, on my way to another forgotten place in my journey where nothing matters anymore, where living or dying is irrelevant, and where your turmoil can only generate the necessary energy to propel you to greater and better things.

Joseph, the son of Jacob, crossed my path. He was the exact picture of his father and had also mastered the same wisdom as old Jacob. He knew of me, as I found out later, from his father. Our journey together was peaceful, and brought me one step further from where Jacob had left me. We shared our lives, our experiences, our hurts, and our joys. We laughed at the world and ourselves for all the serious and laughable events and actions we had been involved in. We studied the purpose of our lives, the reasons of this world's existence, and our relationships with our fathers.

After much searching and licking our wounds, Joseph and I discovered (with the help of his beautiful and powerful bear) the true meaning of success for our passage onto this world:

> To laugh often and love much; to win the respect of intelligent persons and the affection of children; to earn the approbation of honest critics and endure the betrayal of false friends; to appreciate beauty; to find the best in others; to give one's self; to leave the world a bit better, whether by a healthy child, a garden patch, or a redeemed social condition; to have played and laughed with enthusiasm and sung with exultation; to know even one life breathed easier because you have lived; this is success.

Joseph and I rejoiced at this wonderful and most powerful discovery, and departed with warmth in our hearts and in our soul.

As for me, my journey has never been the same. I've always kept these two wonderful lessons present in my mind and in my guts even though sometimes the environment challenges their realities and my obeying their teachings. May these bring you as much wonder and grounding as they have me.

The Price of Freedom

HUNG LE
♂ *EBNOM*

Vietnam, my country, fell apart when the Communists took over in 1975, and my family fell apart as well. Everybody went away. The government took all our property — houses, land, everything. They put my father in jail for 12 years. I had to leave my mother and go to another city in order to avoid the Communists. They were looking for young men to go into the jungle and build new cities. Some time before, our family had tried to live in the jungle for six months. We couldn't do it; it was very tough for city people to survive there. I knew that if the Communists caught me they would send me to the jungle, and I would die.

I didn't see any future for myself at all. I was 14 years old.

I did a lot of different kinds of work just to survive. But I also wanted to save money so I could escape from Vietnam. I had nothing to stay there for. On one of my escape attempts I got caught and sent to jail for six months. After I was released, I could no longer get in touch with my family, because if I contacted them it would mean trouble for them.

In those days, you paid your money in advance to the boat owner for passage out of the country. But you didn't carry any money with you. The odds were 90% that you would be killed and 10% that you would escape successfully.

The next time I got caught escaping they sent me to jail for a year. They moved me from one jail to another but there was no talk of releasing me. So I had to escape.

I was out on a work party with three younger boys. We had to make our escape that day because we knew the prison authorities were about to

move us to another camp far from the city. And far away from the city we would be goners.

Late in the afternoon we fled into the jungle. It was pitch dark when we came to the banks of a big river. We had to cross this river to get to the city, which was somewhere on the other side. There was no boat, no bridge, nothing. We were stuck. We couldn't return to the camp. I decided that we had to swim across the river.

Of the three boys with me, two knew how to swim and one did not. We couldn't just leave the boy who didn't know how to swim. He would be recaptured and taken back to the camp, and he probably would not survive.

We had seen what happened to prisoners who were caught after escaping. They were put in the *caso*. These were metal cargo containers — not the big long ones, but small ones, about 8x8x8 feet. Sometimes the prisoners were alone and sometimes with one or two other people. These *casos* were placed outside, completely exposed to the elements. The climate in Vietnam is very hot, and the sun is fierce. A man in one of those *casos* would bake like a chicken in an oven ... And the nights could be rough as well; sometimes it was freezing cold.

So we had to bring that boy with us. The distance to the opposite shore was several hundred yards; it was hard to tell exactly because the night was so dark.

We started swimming. I swam using two legs and one arm; with my other arm I held the shirt collar of the boy who couldn't swim, and I pulled him along in the water. The three of us who could swim had taken our clothes off, and he was holding them in a bunch. The two boys who could swim were paddling beside me.

Things went well at the beginning. But about halfway across we all began to get really tired. One of the other boys panicked. He started calling my name, and the next thing I knew I felt his hand on my shoulder. So now I was pulling two people.

After about 15 minutes of this I was exhausted; I had nothing left. I was swimming very slowly. The shore was still a long ways away. The other boys started to scream. They thought we were all going to die. My only thought was to keep struggling, keep moving forward.

Finally the people on the other side heard us and sent out a little boat and picked us up and took us to dry land. When it got light the next morning I saw that, while we had been drifting downstream, the river had been getting wider and wider as it approached the sea. And the current was getting more turbulent as it began to run into the ocean breakers. We never would have made it on our own.

Unfortunately, the people who rescued us were working for the government in this little village. They were like security guards; they had guns and everything. They were going to hand us over to the police the next day, and we would be shipped back to the prison camp. I knew I'd have very little chance of escaping again from prison; I'd probably be shipped to Kampuchea or somewhere.

The next morning we woke up before everybody else; they were sleeping on the ground around us. The four of us just got up and walked away. We walked into the village; no one was up yet. We couldn't tell which way to go to get to the city. Dogs began to bark so we just took off running down a road. After a mile or two we came to the ocean, and we walked along the beach. Eventually we came to a section of the beach that was crowded with swimmers and sunbathers. I knew

then that we were close to the city and that it was Saturday or Sunday. We stripped off most of our clothes and melted into the crowd. We stayed there into the afternoon. We saw police cars driving up and down the highway that ran along the beach; they were looking for us.

The youngest of the boys I was with — he was eleven — met a friend of his on the beach. He convinced him to go to our guy's house and tell his mother that he had escaped from prison and that he needed clothes and money for four people right away. We changed into the new clothes, left the beach and walked to a pagoda in the nearby hills that housed an enormous statue of the Buddha. We stayed there all night. The next morning we walked down to the train station and caught a train into the city.

I worked in Saigon for two years, saving money for another attempt to flee the country. By this time I had learned something about escaping. I carefully chose the particular boat owner I wanted to go with. This trip was very expensive; I was able to come up with half the money, and I borrowed the other half. But the odds were better on this boat: there was only a 75% chance that you would die while escaping. And if I got caught, they said they would bail me out. The people running this operation had bribed the right officials. When people got shot while trying to escape, it usually meant that there had been a disagreement about how the bribe money was going to be divided up among the government officials.

Everyone understood that when you made the decision to leave the country, you accepted the fact that the attempt might well cost you your life. It was freedom or death. So I paid my money, and once again I was waiting on the beach for the boat to materialize out of the darkness. This was my seventh attempt to escape Vietnam. I had been captured twice. Three times I had been waiting, but the boat had never shown up. And once the boat had broken down and had returned to shore.

METAMORPHOSIS

But this time, after more than three years of trying, I was lucky. I got out.

I ended up in a refugee camp in Indonesia where I lived like a dog for a year and a half. The Indonesian police were animals, and they treated us like animals. From their point of view we were like renters who weren't paying the rent; we were homeless bums. The living conditions and the treatment we received there were brutal.

Because I had relatives in America, and because my father had served with the U.S. military, the refugee organizations wanted to send me to America. But I didn't want to go. America was the reason that my father was in jail, that my family was split apart, that I was having to pay for my freedom with my life. America had come to my country and hadn't stayed to finish the job; my country and my life was destroyed. I wanted to go instead to Australia or Canada.

Nevertheless, it was up to the U.S. to make a decision about me first. I had been in the refugee camp for six months before Immigration requested proof that my father had been a soldier for the U.S. I wrote my brother in America to help me establish proof; it was another six months before I was accepted to go to America.

I was required to spend six more months in the refugee camp studying English. My teacher was an Indonesian who had gone to college in the U.S. A lot of the English he taught I had already learned on the streets of Saigon. After six months of language lessons I had

to wait two more months before I finally got on the airplane. I flew to Singapore, Hong Kong, Seattle, and finally to Syracuse, New York, where my brother lived. There I began my new life.

My first sight of America was flying into Seattle at night. I looked down at the freeway, and I was amazed. I'd never seen anything like it. It looked like an army of ants moving up and down a path. At that moment I knew that I was free. I was absolutely certain. It's hard to describe the feeling. I felt that a great weight had been lifted off my life. I no longer had to keep looking back over my shoulder. No one would come up from behind me and grab me by the collar and throw me into jail.

If you have never had your freedom taken away from you, it's hard to know what it's worth. I know the value of freedom. I have paid the price of freedom.

ORIGINAL PAINTING BY
PETE MOKRIS
Victory ♂ *SBNoM*

Years of Promise

RICK HERBERT
Golden Gate Jerks ♂ *SFT*

We made our entrances into this world via Middlefield Drive, situated on former sand dunes out in the western end of San Francisco, the last major area of the city to be developed. Before WWII, the area west of Nineteenth Avenue all the way to the Pacific Ocean was sparsely populated sand dunes, as it had been since time immemorial. After the war, a handful of hungry developers leveled the sand and began laying out the straight blocks, with somewhat pretentious names like Sylvan, Meadowbrook, Inverness, Middlefield and Forestview, considering the sandy constitution of the surroundings. The neatly laid-out, compact houses cost our parents about $6000 in the middle 1940's, houses that command upwards of $200,000 in the early 1990's. My brother has a photograph of our mother standing in front of the partially completed house, beaming proudly at this major acquisition that would be our refuge for the next twenty years.

Our paths, our sources ... from everywhere ... from Ireland, Germany, England, Scotland and countless other places. First, second, third generation Americans, native San Franciscans, cab drivers, sheet metal workers, bankers, plumbers, doctors, clowns, drunks, kids from Tennessee with alien accents, Junior Sports Stars, pharmacists, podiatrists, bashful kids, brazen kids, kids with a certain genetic need for adventure, for testing the edges of the acceptable. I'm sure our parents spent many a sleepless night wondering if we'd reach adulthood in one piece, as we ran around proving ourselves to ourselves and each other, exploding small but lethal bombs made from unraveled firecrackers, lying in waiting for the Fleishhacker Zoo train to pull through the tunnel, so we could run alongside and hop on for a free ride, galloping rented horses along Ocean Beach, pulling devilish pranks on work-weary and humorless adults. And on rainy Saturdays when we were forced indoors, we would congregate at Ron and Joe's house, and before long, all of these raw little human personalities, together by tacit agreement to share the fears and joys of boyhood, of childhood, would have some grand fantasy cooked up, like a faceless, murderous monster lurking in the basement; we would make ourselves delirious with fear and excitement, daring each other to be the first to brave the darkness and vanquish the monster. One foggy summer afternoon in 1956, upon hearing that one of the younger kids had lost his toy gun, The Junior Detectives was born, its first case to find this kid's gun. I don't recall if we ever found it, but our club subsequently grew to include quite a few of the neighborhood kids. It strikes me now that we were the generation that witnessed the birth of the Age of Television, but we had been fortunate enough to have developed a rich capacity for creativity and imagination before the Great Eye had a chance to claim our minds. I think how different my daughter's growing-up years have been from mine, and then I think of my dad and how he viewed our lives as we grew, compared to his upbringing in a much simpler, rural San Francisco, one of eight thriving kids. But here we were, experiencing together the post-war boom years and the promise they held.

As so many prime-time TV shows include in their nostalgic rendering of those years, the 8mm movie camera was an incredibly valuable recording tool; my dad used his at seemingly every possible opportunity, and in so doing

METAMORPHOSIS

captured so much of the flavor of those years, and even now that essence is only there when the old Revere projector is fired up and the films are projected on a screen; the act of transferring the film to videotape, as I have done for convenience, somehow robs them of their soul. The still-vivid color of that 40-50 year-old Kodachrome gets lost in the transition, and it just isn't the same without the whirring motor of that old projector. So, thanks to my dad and his obtrusive, blinding floodlights and camera, much of the highlights of our early years were captured for posterity: our wonderful birthday parties with those presents and party favors that I still remember; their color, their smell, the feelings of excitement; the yearly vacation trips to Camp Mather. Wherever we went, he had that camera, and a single-minded quest to capture everything that was happening, to steal that fleeting time, bring it home and enjoy those moments again and again. It is almost like the viewing of those events in the warmth of the living room, with mom and we kids, with a highball in his hand, was more of a tangible reality than the original event itself; in the act of recording those events, he sacrificed a certain being-in-the-moment, a stepping out of the scene in order to record it.

My brother Dave has been the Keeper of the Films for some years now, and often when we get together he will set up the projector and select one of these slices of our lives to view. As I watch those images come to life on the screen, long-buried memories combine with them to fill in the blanks of recall. So much of the foundation for who I am and am becoming was laid down in those years of promise, and I get a certain feeling of wholeness and an appreciation of my place on this little finger of earth at the edge of the Western World.

Abuelo y Neito

ADAM SEGAL
Minutemen ♂ EBNoM

Photograph taken while wandering the back streets of Michoacàn, Mexico, in 1985. This grandfather was very proud to have his picture taken with his grandson.

ORIGINAL PHOTOGRAPH BY THE AUTHOR

Seasoned Oak for Willie

W. K. Dolphin
Guest ♂ EBNoM

PROLOGUE

In 1988 I bought land in the mountains of the Shasta-Trinity National Forest, 12 miles from Hayfork, California, 60 miles west of Redding, to develop a retreat center. I wanted to provide secluded accommodations in nature where people could devote themselves to healing, inner development and creative work. The following years saw construction of a lodge, a bunkhouse and a number of tent camps.

Guests began to use "Silver Spring Mountain Retreat." Some came for short stays. Others, like Willie Barnes, came for retreats of six months or more.

Last fall I remodeled the open-air baths on the lip of the canyon. Upon finishing I consulted the *I Ching*, which prophesied a near future marked by *Increase*, "as the earth partakes of the creative power of heaven, forming and bringing forth living things."

And in fact, marvelous things began to happen; among them Willie's letter, which arrived at Thanksgiving, saying *he was coming back!*

I was ecstatic. In my wallet was the snapshot I'd taken of him the year before. It pictures a short powerfully-built man in his early forties, flopped back on the side of a ditch, shovel in hand, his rugged face lit with sweet exhaustion.

He arrived soon with many a tale to tell of his months in the Trinity Alps, and laid $1500 on me. I bought long-needed supplies and two fine batteries for the solar system. Together we laid pipes as we had the year before and made repairs to the Lodge.

There is a *force* in Willie that sees into people's lives and helps them in their development. Over the years it has directed him into experiences that have transformed him from a psychotic suffering from a childhood of abuse into a broad loving human being with great ability to help others in their growth. He is a private person and I can't say more, except to add that it's fascinating to watch this force working through him—we call it *THE ENERGY*. This memoir tells of an assignment this *force* gave me that enriched my life.

THE BUFFETS OF REDDING

In late December the serious snows began. My eleven-year old daughter and I were house-sitting a friend's place about eight miles from the retreat, some 1,000 feet lower in elevation.

For the first few days I was able to keep the road open between the house and the retreat by towing a V-shaped wooden plow behind my truck, a Dodge Ramcharger four-wheel drive with chains on all four tires. I was bringing Willie water every day and laying in firewood for his 90-day fast.

But the snow got ahead of me. I unhitched the plow so the Ram could rush without drag. Although he was loaded with firewood for extra weight and extra traction, after an hour of charging the drifts, backing and charging, backing and charging, advancing about 1000 feet, he high-centered: he got stuck perched up on his transmission, his four hooves flailing helplessly before he slid into the ditch and died.

I hiked the two miles back to my daughter. The snow continued, and it reached a depth where the local plows couldn't function. The Trinity Pines residents were stranded for a week, until they

METAMORPHOSIS

hired a bulldozer that opened the roads to some extent (and ran into *my* plow, then buried, and smashed it to pieces).

I knew Willie was low on food. Lulled by six years of mild winters, neither of us had stocked up properly.

After making arrangements to return my daughter to her mother, I packed groceries in a knapsack and got a ride to the end of the plowed road. My little Nissan was parked there, five miles from the retreat, immobilized by earlier snows. Buried under a rounded drift, she looked like a bun taken fresh from the oven.

The day was bright and sunny. The snowy forest stretched in all directions. A cool blue light glowed in the crevasses. As I set out I was enjoying the trek. But conditions were unfavorable for snow-shoes: powder built up quickly in the webbing and added five or six pounds to each foot. I would advance 50 paces and rest, letting my heartbeat drop back to normal. In the late afternoon I stopped in a sunny spot on the trail, curled up on the snowshoes and took a nap.

When night fell I was still a good mile from the retreat. I was tiring. I took a shortcut through the forest and began to feel really alone. What if something happened? No one would find me there. I was down to 30 paces between rests. Trees had fallen across the trail, and in struggling through them I sometimes toppled over. Righting myself was something of an art, as the powder offered nothing solid to push against.

But the night was incredibly lovely. A full moon shone down on the snowy forest, giving it the look of an enchanted land in a fairy tale. And so peaceful. Silent except for a mild whispering wind. I crossed the tracks of rabbit, bobcat and deer. The smaller trees were bent double by the snow, arching over in beautifully sculptured shapes. I thought of the poem by Robert Frost, "*The woods*

are quiet, dark and deep/ But I have promises to keep/ And miles to go before I sleep."

The last half mile took some grit. Uphill. 20 paces between rests. I was grateful that no human hostility added to the situation, no enemies crouching in the next grove of trees, Serbs spring-ing out to aim and fire. The thought of a cup of hot chocolate by Willie's fire kept me going.

I reached the gate at last. As I trudged by Shasta Bunkhouse, I saw that its roof had collapsed. The four walls were canted out about a foot, and the roof — a shed roof, the last one I'll ever build — lay crushed to head level under a good five feet of snow.

Willie and I celebrated with a big dinner. He hadn't been sure ... of a lot of things, including whether he was going to make it. He had no skis or snowshoes. Even walking to the outhouse had been a near impossibility.

Before Willie came to the retreat he had troughed up at the buffets of Redding, putting on an extra 60 pounds to prepare for his fast. Daily for three weeks he would arrive at one or another buffet at 11 a.m. and eat and read, off and on, until 3:00 when the lunch buffet closed. But now, as he carried all that extra weight through the dry powder, his every step plunged him into the snow up to his neck.

The following morning Willie revealed that the firewood I'd laid in was insufficiently seasoned. It burned without heat. Moreover, it was nearly gone.

A friend had very kindly lent me the use of a snowmobile for the winter. It was at her summer cabin a quarter mile away. As I planned to spend the winter at the retreat myself, finishing my novel, I proposed to Willie I bring him my propane heater and use this snowmobile to ferry him tanks of propane during the months ahead.

He stiffened. His face hardened and he declared emphatically, "NO!"

This was not Willie, not at all. *THE ENERGY* was speaking. Through him, *IT* announced that I was to remove Willie from the retreat and provide him with *three cords of seasoned oak* and one of fir *by the next full moon*. Or he would not come back.

An ultimatum! What a thunderclap! Willie was astonished and apologetic. But he was long accustomed to acting on *ITS* orders and felt he had to comply. He thought I might benefit if I too complied.

I considered. Willie was one of my best friends. I was actually beginning to think of him as my teacher. He was magnanimous; he had money; he'd been practically my sole support last year. And he had possible access to some major funding down the line.

Then too, he loved Silver Spring Mountain Retreat. He had a singular connection with the place. He'd poured his own sweat and blood into helping it grow. And he wanted to finish his process here.

I decided to accept the task. The obstacles were great: the road conditions, the scarcity of dry seasoned oak in this season, the exertion. Nevertheless ... if it could be done I'd do it. If it couldn't, I'd have done my damnedest.

We discussed what the job would take in the way of money. In this area, seasoned oak in summer goes for $85 a cord, fir for $60. Considering the present demand, and the distance from town, we decided to double that figure. Adding two extra cords for my use, we came up with a figure of $900 for the wood.

In addition, the road from the Nissan to the retreat would have to be cleared of snow by a plow. I figured this could take as much as a whole day, which at $65 an hour would come to around $500.

So ... $1400.

Willie had $900 available now, with more to come later. I would pay for my two cords myself.

Could I talk a bulldozer man into taking the job? We decided that in this impoverished area a display of cold cash would sweeten the possibility. As an environmentalist I was nervous about approaching the macho logger who owned the bulldozer down in Trinity Pines. We role-played the exchange, Willie as the logger and me, at the critical moment, flashing a wad of hundred dollar bills.

That night I did an *I Ching* reading which foretold, "effort, will power, and movement that will not be worn down by hindrances."

Now began a month of toil which, paradoxically, yielded an inner peace through coping with the world ... a peace that I hope my guests will experience in retreat.

To begin with, I had to get Willie off the mountain and to Redding where he would wait in a motel while I delivered the wood. But the roof of my cabin was sagging ominously, and we first had to spend a day shoveling snow off the buildings and trailers. Then I snowshoed up to my friend's cabin and dug into the mounds where two snowmobiles nestled side by side under tarpaulins like ponies in a shed. Neither of them had been run for a year. "Bucky" and "Matilda" I named them, and following the operator's manual, I performed maintenance procedures on them like a veterinarian taking care of valuable neglected animals. By dusk Matilda kicked into life.

That night I fell prey to a fear I couldn't name. It was paralyzing. Willie advised me to acknowledge and accept it. He counseled me to welcome all emotions as the substance of life while cultivating an awareness of the timeless inner being that generates the self, beyond the emotions, lifetime after lifetime. And he gave me two books on mindfulness. I lived with these books night and day throughout this month of toil. The practices they contained led me to an inner peace as I struggled with my task.

The next day I was amazed and inspired by the sight of my snowshoe tracks from two days before. *Someone* had trudged six miles up the mountain to deal with this situation. If that someone could do that, I could surely take over.

A second day of work on the snowmobiles ended with Matilda's actually surging off her trailer into the snow where she rumbled in place, digging her way down to bare earth.

A couple of days of sun had turned the powder wet. Could she function in these conditions? Would she actually transport the two of us on her back? The steps ahead stretched on endlessly, each dependent on the next. Get her down to the retreat to pick up Willie, get us to the Nissan, shovel the Nissan out and take Willie to Redding, find a place to live in Hayfork while organizing the wood effort, find the wood, plow the road, deliver the wood, ferry Willie back to the retreat ... the prospect of all this sent an army of ants swarming through my gut.

Fear returned that night in full force. I found relief in the books. They declared that Mind lies at the core of being, Mind which, if unperturbed, composes itself like still water, calm and clear, reflecting all things without prejudice. The way to reach this calm state is through paying attention to one's breathing. Focusing on the taking in of air, the release of air, draws one away from past regrets and future anxieties, and into the Timeless Now.

The next day I devoted to Bucky. Since Willie now weighed two hundred pounds, our chances of getting off the mountain would be better with two steeds. But try as I would I couldn't keep Bucky running. Ministering to him took endless exacting procedures. When I was tempted to take short-cuts, I thought of Don Juan, the central figure in

Castaneda's books, and his advice to Carlos that a warrior leaves nothing to chance.

And I prayed to the spirit of Silver Spring Mountain. I asked her to help these ponies run. Willie and I had talked about the power of Mount Shasta. She is widely regarded as a sacred mountain. A live volcano in a dormant state, Shasta contains a core of white hot magma, a fiery jewel that broadcasts a nourishing energy to the surrounding country. So we believed, at any rate. Was it possible that such energy lay dormant in all mountains? Could it be activated and directed? Meditating together we focused on the heart of our mountain, asking for its help in furthering the retreat and in relieving suffering everywhere, particularly in Bosnia and Somalia.

By dusk it had become clear that Bucky's problem lay beyond my expertise. I returned to Matilda, shoveling a ramp out in front of her which I hoped she could climb. One step at a time, the books advised, and each step performed with one's full attention, to the best of one's ability, believing in success.

As I worked I took pains to surround the performer of the steps with loving care, eating well, getting plenty of rest, giving myself encouragement and support.

During a rest period an insight arose. I was sitting by the fire, warm, nursing a cup of cocoa, when an anxiety attack came on. *Fear anticipates what's ahead.* It anticipates freezing hands and feet. It anticipates unrelenting exertion in the dark. It anticipates uncertainties without number. Yet once one is out in the elements, preoccupied with a task, one is hardly aware of discomfort.

I relaxed into the warmth of the fire, the taste of the drink.

As the temperature dropped, the snow froze, and Matilda acquired the traction to crawl out of her trough. These beasts are steered partially by the handlebars and partly by one's weight. She was hard to control on the drifted sloping trails skirting the mountain. Again and again she spilled me off and died, stuck, and had to be shoveled out and restarted with endless tuggings on her tether.

Willie and I had to move before the sun unfroze the snow and turned it wet. We rose before dawn and trudged out to Matilda, got her turned around and started. And lo and behold, she performed marvelously! Willie's extra weight was *no problem!* She zipped effortlessly down the trails. It was exhilarating! We whooped with glee! The only problem was a big oak lying across the road. The little coping saw I'd brought couldn't hack it. We ended up digging a tunnel under the trunk and driving Matilda under.

In the early afternoon we reached the Nissan, hungry. As we shoveled her out Willie held forth on the buffets of Redding. I learned about their entrees, their salad dressing, their desserts, their clientele, their decor. All this so I could make an informed choice as to which one he would invite me to once when we got to Redding.

By dark we'd shoveled her out. Only a quarter mile of partially plowed road stood between us and the highway. Unable to turn the car around we made successive runs backwards that yielded about six inches per run. At 2 a.m., just short of our goal, we high-centered and, exhausted, slept in the car and shoveled her out in the morning.

I drove Willie to Redding, arriving at 3:15 p.m., too late for the lunch buffet. Leaving him at a motel, I returned to Hayfork to begin the firewood effort. So ended the brute exertion — for a time.

☙

THE WOOD

First I had to establish a base of operations. I looked for a place to rent. No luck at first. One man was so hostile I had to laugh. "Hell, no!" he growled. "Move! You're blocking my truck!"

Friends directed me to a bar that used to rent rooms. The proprietor was a white-haired soft-spoken man known for helping people out. I explained my situation, and he showed me to a corner room over the bar and checked that the wall heater worked. The floor sagged and the door wouldn't shut, but the room was clean. It contained an iron bed, a chair, and a battered dresser scorched with cigarette burns. Down the hall was a bathroom and shower. Luxury! The water system had frozen at the retreat; we'd been melting snow for water. I asked how much the room would be, but he was noncommittal. This was to be my home for the next two weeks.

Looking into the wood situation, I found that this storm, coming as it did after six years of drought, had caught most people unprepared. Firewood was in great demand. Prices were high, often double. About a dozen woodcutters were actively supplying the town with such snags as they were able to get to. Much of it was fine seasoned fir. No one was selling oak, though, let alone *seasoned* oak, which requires six months to dry.

And seasoned oak it had to be. *THE ENERGY* had been emphatic about that.

I called Hugh, the bulldozer operator. He told me to meet him at the bar. He lurched in wearing red suspenders and a baseball cap. Man in his late forties, barrel-chested, kneading his paunch. Everyone along the bar greeted him — a celebrity, one of the town movers. He plonked himself down on a stool and squinted at me with ragged green eyes. One eye gazed off at an angle. "I'll plow your road," he announced in a high rasping voice, "but I'm telling ya I'm not into bullshitting around."

It was not a matter of talking Hugh into it. Far from it! He wanted to plow that very day! Tomorrow at the latest, because he was going to Tahoe over the weekend. I told him that first I needed to purchase the wood and line up transportation.

METAMORPHOSIS

This seemed to be the time to flash the hundreds. I did.

"That don't mean nothing." He rocked back, offended. "I've got money in my pocket too."

No, he wanted to plow today and if it kept on snowing he'd plow the son-of-a-bitch again tomorrow, that's all. What was the fucking problem?

I said I didn't want to pay to plow it more than once.

"Jesus Christ," he snorted. He'd plow the son-of-a-bitch for nothing. Yes, he would. Just to get it done. He glared at me and said he had a mind to go up and starting plowing the god-damn thing right now.

"Please don't," I said. "I'll let you know when I'm ready."

Later I came back to the bar and spotted a good-looking woman who'd agreed to lend me her truck. Facing her was a semi-circle of local dudes, including Hugh on a barstool nursing a mug of beer. As I slid in beside her, he squealed like a boar and hit me in the chest with a bottle cap, to a chorus of guffaws.

That day I met Bob Riley, a woodcutter who stressed his reliability. He assured me he could fill my whole order. Deliver it too. He even had oak! I went over to his place to see it. There in his shed was an actual half cord. He was going out for more that very day! "Excellent," I said. "I'll be back in the evening to pay."

"Do you want some oak and fir mixed?" he asked as he pulled out.

"No," I said. "Concentrate on the oak."

I came back that evening to find him stacking firewood into his truck. The beginnings of my order, I assumed. Wrong. He'd brought back fir,

and since I didn't want mixed, he was combining the half cord of oak with the fir to make up a mixed cord he had sold to another party!

I combed the town for leads. A friend told me of a man who had lots of wood in his yard. I went to him with my tale. Short snappy fellow in his mid-fifties, hardy, self-possessed, wearing Stetson and lambswool jacket. Mound after mound of seasoned oak stood along his fence in neat one-cord stacks covered with tarps. Stable out back. Next to the ranchhouse an aviary of carrier pigeons.

I explained that my livelihood was at stake. He informed me he wasn't in the wood business, we were looking at his personal supply. But he'd think it over. Come back the next day. I did and he sold me a cord of oak that had been seasoned for two years for $90!

Transporting it would be up to me. I went to see a tall monkish fellow by the name of Owen whose 53 flatbed stood mired in drifts behind his mobile home. He couldn't budge the thing without chains. He offered to deliver my wood in exchange for a set. They had to be double-width to fit the tandem wheels, and the tires were an odd size.

What a deal...if such a set could be found.

At a heavy-equipment yard a man crawled out from under a backhoe and lent me a set on the spot, refusing a deposit. This got Owen's truck on the road.

Tandem chains, new, went for $200 and would take two weeks to deliver. Owen suggested I visit a millwright known for his resourcefulness. Broad-shouldered and muscular, in greasy overalls, this man had the hard look of a convict. His small slanting eyes contrasted startlingly with his friendly manner. With a welding torch he set to work combining scraps hanging from spikes in his barn.

The task was chains, and I gave him my full attention. As we toured his yard he told me his life story. I learned where he'd bought his various pieces of equipment, what he'd paid, how he'd rebuilt the hydraulics on this one, how someone had stolen the booster box off that one, and how he'd gotten it back. How after a drunken brawl the police had come to the house to confiscate his guns, but he'd had them hidden. We laughed. Pals, though strangers.

The chains he pieced together looked almost new and they fit perfectly. He charged me $60, which just might have covered his time.

I phoned Hugh to tell him the order was building and found him astonishingly *accommodating*. He was in bed with a "whiskey" flu. "All right, Dolphin," he whispered. "Let me know when you're ready."

Searching for seasoned oak brought me in contact with every woodcutter in town. Most of them belonged to a loose society of out-at-the-elbow young men who dredged for gold, cleared brush, worked on cars and cut firewood depending on the season. Their day began with beers at the bar and a borrowing back and forth of saws, wedges, mauls, and the inevitable $5 for gas, one collecting a dollar or two from another to lend to a third in a complex exchange in which everybody seemed to owe somebody something. In the evening they drifted back, usually with money enough for the day's food, some billiard games, and their bar tab.

They promised to look for oak snags which had been standing dead for some time and were therefore seasoned, but rarely were they successful. I bought such smatterings as came in, a quarter cord one day, an eighth of a cord the next. For the most part they were good-hearted boys, outlaw types with chewed-down nails and jaunty caps, smoking pot with short fierce drags.

But things didn't always go as planned. I paid for a cord of fir and suddenly it vanished. The woodcut-

ter, a kid who chewed tobacco and wore a stud in left ear, had sold it a second time to somebody else! "Don't worry, buddy," he patted me on the back. "I'll get you some more tomorrow." Which he did.

Evenings I sequestered myself in my room and poured over the books. Turning on the lamp wasn't necessary. The room was awash with yellow light from the sign outside the window. *PLAY LOTTO HERE* it urged, and announced the upcoming Miner's Ball, when Dirty Dick and the Grinders would blast away in the Cinderella Ballroom.

One night a commotion in the street drew me to the window. Pulling back the curtain I discovered there was no glass in the frame. No broken glass on the floor, either. I realized the pane had been missing all along. Yet the room was always toasty. The stove that heated the bar, I discovered, was directly under the floor.

The books seemed to speak to my situation. One passage struck me particularly. *Be at ease. Be as natural and spacious as possible. Slip quietly out of the noose of your habitual anxious self. Let peace gather your scattered mind into the mindfulness of Calm Abiding.*

Calm Abiding! The phrase thundered in my head. It evoked an ocean of peace at the heart of being. Again and again I returned to it, particularly while working on my task.

Even while I was producing *effort, will power, and movement not worn down by hindrances.*

Yes, the task *was* requiring effort, not the brute exertion of before but definitely an all-out persistence. And a pattern seemed to be developing. Most people seemed only too willing to help. This room, the cord of oak, Owen's truck, the chains, Hugh's new flexibility — all had come easily. I wondered whether Willie was right. Had *THE ENERGY* foreseen something valuable for me in

METAMORPHOSIS

this task? Something beyond mere economic survival? Had *IT* possibly assigned the task to me, to bring me through some stage of growth?

Or was I simply going about things differently, in a way that elicited this positive response from people? Well, yes, I was. I was assuming full responsibility for the job. Giving whatever it took ... the pattern began to make sense. One of my life issues has been self-reliance. From adolescence through early adulthood, even into my middle years, I required the energy of others — to live. Required their endorsement, their love, their encouragement — to function. If I'd taken on this task at an earlier time I might have begged, "Look what I'm up against. It's too much. Can't you see I need help?"

Yes ... self-reliance. But then what is the self? And what can be relied on? One's personality? One's values? One's character? This experience seemed to be pointing to an inner force deeper than all these.

My thinking was often derailed by raucous sounds booming up from the bar. Nightly a rollicking crew gathered there. Doors slammed, shouts broke out, arguments raged. Sometimes the cops were called. (A man was actually killed there in a brawl a few years back.) The billiard players punctuated their shots with hoots and curses. And over and over the juke box bellowed the tale of a girl who wore an itsy-bitsy teeny-weenie yellow polka dot bikini.

A mechanic and I spent a day up on the mountain reviving the Ram and shoveling him out of the ditch. Countless trips to a distant community turned up half a cord of madrone, an oak equivalent. The order was coming together, though spread out all over town. It was time to think about plowing. The deadline was approaching. I was still short a cord of oak, however. Swallowing my pride, I went back to the man who'd sold me the first cord and asked him for a second — and he obliged!

§

THE PLOWING

The flatbed had to go in right after the dozer, otherwise new snow could make the road impassable again. The weather pattern had been four days of snow, two of gray brooding sky, then four more of snow. Would it hold? The forecasts were absurdly unreliable.

I watched the sky, and when stars appeared over the mountain, I gave Hugh the go-ahead. This set in motion a train of events that couldn't be turned back. Snow promptly started the next morning, but fortunately the cat broke down, and we laid over a day for repairs. The day that followed was clear.

Plowing the road proved to be a social affair with a flavor all its own. Hugh brought his wife onto the scene, and three sons, strapping hard-drinking lads in their early twenties, and their pack of hell-raising buddies. The cat was a D-6 with a ten-foot blade and an oversize diesel engine, a bucking whopping Tyrannosaurus of clanking steel. It roared with raw power. Toby, Hugh's eldest, was its devoted keeper. Perched on its armored back, he directed its rage.

Hugh was his old self again, and he strode around like Patton, barking orders. The brothers pumped blood into the fuel tank. Axel, an old lame bulldog of a man, was to follow in a pickup in case of emergencies. With my Ram on the scene, we now had a cluster of small trucks surrounding the beast as it drank and prepared to eat its way through five miles of snow to the retreat.

I was hoping Toby could open the road in two passes, one up and one back, but the treads spun when he took too big a bite. Five passes were going to be necessary. My heart sank. What would this mean at $65 an hour? I'd gotten good prices on the wood. Willie and I had figured on $900 but I'd done it for $650 with only a cord of fir still to go. But it now looked like part of the plowing would have to come out of my pocket, money I'd have to borrow. I asked Hugh how long he thought it would take.

Hugh shot me a satisfied smile. "This is going to cost ya."

I breathed deeply. This in fact was the high point of the experience, this morning that I spent entirely in the present, following my breath, well-rested, well-fed, no pain anywhere in the body, no anxiousness in the gut, serene in mind and without thought to the next hour. I was at home in homelessness, in my heartbeat, in mindfulness, in Calm Abiding, enjoying the snowy countryside and the unfolding of this action with its wily cast of characters, each unique, each stirring my affection, and thinking fondly of Willie waiting in Redding, feeding at the buffets, knowing I was getting him his oak. I was the one causing this to happen, yet an intrigued spectator at the same time, a witness to the timeless sun rising over the mountain, a witness to the universe's need to express and understand itself, to form in ever changing patterns and take itself by surprise.

Toby plowed a mile at a time, up and back, up and back, with extra work at the turns. I rode with Axel, the old dog, and listened to his tales of times past. He used to run a big logging outfit in the Sixties. He told me of snows so deep a cat ran over a buried logging truck without even noticing. No damage was done to either machine. He told me of how distracted he was on learning of a minor accident that befell his daughter that he failed to notice a log rolling off a truck that flattened and nearly killed him. He told me of his recent bout with thieves trying to steal a loader out of the forest late at night, a story involving guns and ending in death. He often paused to fix me with a keen eye and cluck, "You know what I'm saying."

The sun went down but Toby bore on relentlessly, pausing only long enough for his brothers to pour fresh blood into the beast. He rode exposed in

parka and bear-like mittens, never complaining, bucked and jounced by his beast. The temperature dropped and a mounting wind whipped horsetails of snow across the road, concealing him from view.

Darkness turned Axel and me into anonymous males on watch. He went on to hunting tales, never doubting my approval, recounting how he treed and shot a female bear out of season — a momma bear, he called her with strange affection — and how in gutting her he came upon the cubs, curious little teddies that he dug out and pickled and gave to the grammar school for display. In another tale three bucks were resting in the shade near my retreat. At the sight of him they sprang up and fled. He got off some shots. One of them hit the big one in the bunghole and paralyzed its hindquarters. It tried to escape, scrambling with its front legs alone. The bullet went out its shoulder and hit the second buck. Another wild shot brought down the third.

"Three at once! You know what I'm saying!"

At this I left the truck and rode with Toby.

I climbed up on freezing steel and clung as the monster crashed through the oak Willie and I had tunneled under. Talk was impossible over the roar, but Toby often nudged me, passing his whiskey bottle — Black Velvet.

Then Hugh spelled Toby, leaping up on the cat and whipping it through its paces like a bumper car at a county fair. Sheets of snow flew in all directions. Trees rocked, and roiling mounds of snow tumbled end over end to burst out over embankments.

Three hours after dark we reached the retreat, and by eleven that night we'd plowed all the way to the Lodge and back in a double pass. Total time, thirteen hours ... $850.

The next day the flatbed crawled in behind us with six extra hands on top. The plan was to unload fast so the truck could zip back to town for the second three cords. I figured two more hours of plowing would finish the job, but in fact it took seven because of the innumerable passes that were necessary, everyone cold and waiting, and even then we didn't reach the Lodge.

This was the low point of the experience. I couldn't summon the least echo of Calm Abiding as I clung in pure misery to the beast, shivering with cold and deafened by the roar. Endlessly, endlessly it rammed mountains of snow into pockets along the road, upending my beautiful trees, crushing plywood, mangling a solar heater, $1200 into the job and the clock still running.

Even then the flatbed couldn't manage, and the cat had to tow it in. In twenty minutes eight people who'd shivered for hours in the damp cold dumped the wood on the volleyball court 500 feet from the Lodge and headed for home. It was past midnight. A second run by the flatbed was out of the question — then, the next day or ever. The flatbed ran out of gas on the way back from hours of idling to keep the heater running.

In the morning I borrowed a large sum of money and deposited it to my account. Checkbook in hand, I walked over to the bar and slid onto a stool next to Hugh. We agreed on the total hours. Men up and down the bar watched us in the mirror.

Hugh looked away thoughtfully, toying with his pencil. "I'm knocking an hour off."

I stared at him in surprise.

"Is that all right? You want to argue about it?"

"No, but ... I mean, why?"

He turned and looked at me. "Because I want to, Dolphin," he said softly. "That's why."

I thanked him profusely and wrote out a check for the reduced amount: $1335. Still a grievous

gouge. I said Toby had done an excellent job. He had. The damage he'd done was inevitable.

Hugh endorsed the check and handed it to the proprietor who took it to his office and came back with cash. Hugh peeled four hundreds off the stack and gave it back to the proprietor. Since the controversy over the spotted owl, the local economy had been going from bad to worse. The other men came over. As Hugh doled out money, I witnessed a complex swapping around of sums like the exchanges I'd seen among the woodcutters, but with twenties and fifties, everybody seeming to owe somebody something, Hugh no less than the rest.

I thought of the trouble I'd be creating if my check bounced.

Hugh wouldn't let me go without buying me a drink. He grew mellow, confiding. What kind of a deal did I have going up there anyway? I didn't think he'd get it so I told him Willie was a writer, like myself. What did I write? Fiction. He drew me out on my novel.

"I read a lot of novels," he observed. "Not the love-me suck-me stuff. Naw, I like a good story. Mark Twain. Stephen King. Man in my kind of business — I know it don't amount to much — but I've got a lot on my mind. Helps me go to sleep."

I tried to picture Hugh in bed reading.

"There was a day ... but now, with all this bullshit ... rules ... we're thinking about pullin' up stakes ... if it weren't for my boys ... but I gotta keep em employed ..."

I was enjoying this, but I still had three cords to move. I stood up and thanked him again. We shook hands and hovered there uncertainly for a moment. I genuinely liked Hugh now and could have embraced him, but I knew he'd balk at that. We shook again and he smiled and said gently, "All right, Dolphin."

I found the proprietor and asked for my bill. The good man thought for a moment. "Six dollars a night."

❧

BREAKTHROUGH

By noon I'd found two pickups willing to deliver two cords for $25 apiece. The drivers were so disgusted by the punishment their vehicles took that neither of them would return for the last cord.

Still and all, five cords — *five cords!* — in a huge pile. Willie could come back! As for the sixth, mine ... somehow I'd manage.

My spirits began to soar. I broke out a bottle, put rock music on the boom box and began to jig and holler. I was dancing in my kitchen, preparing dinner, when a flashlight appeared at the window.

Who in hell ...?

It was Toby, as sweet a fellow as ever walked this earth, announcing, "I've brung your sixth cord."

Directly, a pickup truck, blaring music, came blasting up the road. Another followed hard on its tail. The brothers jumped out, and dogs, and the crew, and snatched logs from the back and flung them rapid-fire onto the pile. *Pow!-pow!-pow!* the wood flew like machine-gun bullets. In eight minutes they'd emptied the first truck. Delighted and amazed, a little inebriated, I tilted my head back and whooped for joy. This puzzled them at first. High-energy action was their mode. But they took it as appreciation and laughed and really applied themselves, making a game of going as fast as they could, shouting themselves, whooping and cheering, a mad happiness sweeping us up, and in less than five minutes they'd emptied the second truck. It was actually more like a cord and a half, and I enjoyed paying Toby a good price in cash.

I invited them in, but no, they had to go. A driver popped his hood to check the oil and they all surged over and peered at the stick. "Load up," they told the dogs. A dog hesitated to make the leap over the closed tailgate. "Load up! Load up!" they insisted, and the dog overcame its fear and sprang in back.

I yipped and howled as they pulled out, and they yipped and howled back to the sound of backfires and smashing bottles and gunshots growing fainter into the night.

A last step remained. I had to move four cords up to the Lodge before I could bring Willie back. How to do this? Toby had estimated four more hours of cat work to properly open the road. No way could I handle that. I estimated that by building a sled for the snowmobile, I could tow the wood up in 126 trips. Or, by shoveling out the road enough to get the Ram through, I could do it in only 16. I chose the latter. It took three days.

I had to beat the Ram practically senseless to force him to haul those loads. But how gratifying to see the oak mounting up in tiers next to the Lodge! On the final trip the motor mounts split and the engine dropped into the frame. The Ram would still inch forward but couldn't back up. I hoped I could crawl him into town, and I winched him around a switchback and started out. But the engine clunked down through the frame to a shower of sparks as the pulley ate into the tie rod.

The hike to my Nissan parked at the highway was 13 miles and took a long time, though it wasn't too hard because we'd plowed the road. I thought when I set out that I wouldn't encounter anyone till the last quarter mile and that's exactly what happened: an acquaintance cruised up, and I waved him by.

I drove to Redding and found Willie and told him the job was done. He hugged me, eyes sparkling, and hurried us to a buffet of my selection. Each of us ate three entrees, half a dozen vegetables and two sundaes as I took him step-by-step through my adventures. He rented me a room in the motel, and I drove him to the retreat the next day.

METAMORPHOSIS

Willie gave me the money he had on hand, $1500, as advance rent; reimbursement for the plowing would come later on. Although I'd planned to spend the winter at the retreat, this month of toil had worn me out, and I wanted a gentler place to finish my novel. But work remained before I could leave: I needed to get the Lodge ready for Willie's fast.

Wearily I rolled up my sleeves ... but *THE ENERGY* stopped me dead. *IT* announced I was to leave, not soon but *right away*; my labors were done, the lesson over, I needed rest before getting into my novel. Oh, *IT* was emphatic, as much as *IT* had been about the oak. Willie again bowed to *THE ENERGY*'s will — is his life his own? — though this would mean hardship. Instead of drawing water from the tap, he'd be melting snow or hauling it by the gallon from the spring.

Willie and I had our last meal together, a gourmet feast with music and champagne. Then we meditated. I prayed to the spirit of Silver Spring Mountain, giving thanks for all the help I'd received. I asked that Willie be loved and sustained in the months ahead when he would be alone here. Alone, a man sweet and strong under sun and moon, dwelling in silence, the splendid silence stirred by the wind that moved through the trees in a whisper, in a roar.

The bobcat and the deer his companions.

Sustain him, I prayed, as he goes on his journey down ... down ... deep inside to the knots that he will have to untie before he reaches the boundless sea of love that dwells in the heart.

EPILOGUE

For me, I knew, a cycle had come to an end. I consulted the *I Ching*. *Breakthrough*, it announced, "where resolute action brings about a change."

What this change would be I couldn't know then. That I would go down to the lowlands. That I would visit my former wife and tell her I wanted to live nearby to see our daughter and finish my book, and that she would invite me to live in her home, our relationship to flow again like a spring after years of drought. That I would be given a studio where I could work alone, undisturbed, rent free. That the main character of my novel would jell at last, and the complex ending work itself out. That I could rest from this difficult work in the warmth of family love.

None of these gratifications would be diminished by word from Willie later on that *he hadn't needed the wood after all!* For, once he began to meditate and fast, *THE ENERGY* manifested itself as Kundalini serpent fire. *IT* warmed his body to the point where, even on days of bitterest cold, he felt he was burning up!

Finally, of course, the taste I had had of the Timeless Now in my month of toil would remain as something to build on. It wouldn't always be fresh or strong; in fact for a time it would feel like it had never happened. But it *would* come back, I would sense it there, down in my breathing, a pure *perception* in awe of the world, in love with life, free of a me I had to promote or had to defend —moving within me as on the first day we plowed.

It would open a new chapter in my practice, to cultivate and stabilize that state as my ground of being. It would call for daily exercises like those in the Course in Miracles, but using material from the teachers I'm most drawn to: Chögyam Trungpa, Sogyal Rinpoche, Thich Nhat Hanh, Stephen Levine, Jack Kornfield, Lao-tzu, and Don Juan.

Such would be the harvest of my task.

Information on Silver Spring Mountain Retreat can be obtained by contacting Ron at Box 207, Hayfork, CA 96041. (916) 628-5489.

Where Is This Place?

VERN MCGEORGE
Victory ♂ SBNoM

This is a real place and you have to work your butt off to get here.

The team was a real mixed bag, everything from virtual mountain goats who hike and fish and hunt this area on a regular basis to men who had never been backpacking before.

The hike in was in two stages. The first day was up 2,000 feet in about two and a half miles and down 1,000 in about four. As a good team, we had strong hikers in front to lead the way and strong hikers in the rear to hold the team together.

The second day was less than three miles, but the middle mile gained almost 2,000 feet. This climb was one of the most exhausting physical ordeals that I have ever encountered — and it was worth it.

After crossing over the ridge, the hike ended with a long slope where we jumped down from rock to rock in order to lose the last 300 feet. When we got to the camp, we were surrounded on the left, rear and right by high rocks. To the right, a spring provided water chilled by its passage through a snow field that had not yet melted even though this was mid-summer. To the front, Mt. Shasta was framed by the rocks and the lake. As the sun went down, the mountain changed from granite blue to a vivid pink. The full moon rose beside it and lit up our camp through the evening.

Where is this place? Out of respect for the man who showed it to me, I will let it remain an uncrowded secret. Besides, it is not really important. This place is anywhere that men work as a team to achieve a difficult goal.

Until the day I die, I will remember this place and how it looked that evening. Until the day I die, I will remember the men who shared this journey with me.

ORIGINAL PENCIL SKETCH BY THE AUTHOR

METAMORPHOSIS

Himalayan Adventure

DAVID COHEN
Cornerstone & SCMC

In February 1991, seven men set out from Santa Cruz, California for the Himalayas. We were five Sterling men plus a father and son who had not done the weekend. We met numerous times before departure to get to know each other, talk about our expectations and fears, and to become a cohesive team. Our lives could depend on it. Most of the men had not been out of the country before, and I knew simply arriving in Katmandu would be a major adventure.

My goals were to create a supportive intimate team, learn about leadership and for everyone to have way too much fun during our adventure.

I have been to Southeast Asia and Nepal numerous times by myself: once with another man from my men's team and once when I led a trek to Nepal with two other Sterling men and my wife, Gayle. I initially went to Asia for adventure and to bring back temple artifacts, mostly from Burma, which I sold to galleries, museums, and collectors. I always finished my trips trekking in Nepal. That was dessert for me.

I flew to Katmandu ahead of everybody this trip to meet our Sherpa guide and to find a cook and 12 porters as well as to set up our accommodations.

Many of the men wanted to see Mt. Everest. I personally didn't care, but it did mean that if we were to see Mt. Everest, we would be

trekking on a well-travelled path for about four days. An important lesson in leadership for me was that a leader takes his men where they want to go. OK, I would do that. When I am by myself, I try to spend as little time on main trails as possible so that I can get into the back country, away from Coke bottles and shit-eating dogs. Well, it took us four days to get to a place where we could jump off the main trail. Four days of going up mountains and down valleys.

The trails are many hundreds of years old, and the local people have set stones to walk on. Stones that are worn from walking. It is as if you are

ORIGINAL PHOTOGRAPHS BY THE AUTHOR

walking up stairs for half a day, then down the other side for the rest of the day. Sometimes it would vary, and we would walk up all day. We would ask Chuldim, our guide, "Well, what do we do tomorrow?" and he would incline his hand and say "Tomorrow we go up; then we go down and, after, we go up." All right! Pass another three-quarter ton chapati!

On the fourth night, we camped in a beautiful valley below an incredible monastery next to a river. We were told by a local herder that if we climbed above the monastery (all day), we could camp and see Mt. Everest, then we could circle back around and return without retracing our path. No more teahouses; only tiny villages off the main trail and exquisite countryside. That's what I like.

We visited the monastery for a day and just hung out in the valley. Next day, we went up through a snow-covered forest, passing yak shacks, eating yak jack, and then set up camp at about 12,000 to 13,000 feet.

We started out with 12 porters. I didn't want our men to carry anything if they didn't want to. The porters carried tents, sleeping bags, mats, kerosene, lanterns, food and cooking equipment, and anything else of ours. One of the reasons I like going with porters and cooks and a guide is that we are traveling with a Nepalese community, and it is fascinating. We got to know all of them. I learned a lot about leadership from Chuldim, who was in charge of all the Nepalese and our safety and comfort. We felt his leadership without being aware that he was doing anything. Another reason to have porters and cooks is that they bring you warm washing water at 6 a.m. and tea or hot chocolate or coffee and cook breakfast, clean up, set up our tents, take down the tents and walk ahead of us to have lunch ready when we arrive. That's trekking!

We sent back a few of the porters at the monastery because our loads were lighter now. And we asked them where they would sleep on the way back, since they didn't have any blankets with them. Oh, they wouldn't sleep; they would be back to the main town in one day if they hustled. This fact was very humbling, almost embarrassing. We would have dropped dead going any faster than four days.

These people are devout Buddhists and we noticed they get genuine satisfaction in serving others. This is not the kind of giving I'm more familiar with, which is more like negotiating. We began to love some of these men that we couldn't even talk to, just hanging out playing with them and travelling together. There were magical times when the porters would be walking and singing and we would run to keep up to be able to be near them and hear them sing. When we stopped, they always had energy to wrestle with each other and joke around. We, in turn, had just enough energy to decide what kind of cookies we wanted with our hot chocolate.

At camp, above the monastery, it was cold. We went to bed after playing music with the porters in their tent. Chuldim had his own tent that he shared with the cook.

During the night it snowed. The chilling wind, blowing at perhaps 50 miles an hour, blew away the porters' tent. They slept like puppies all curled up next to each other to stay warm with a few thin blankets. The wind wakened us early to see everything covered in snow. Too cloudy to see Mt. Everest. And no porters.

During the night, the porters had split, gone down the mountain to hang out at the last village we had passed, to await our decision whether to continue and in what direction. We hung out at camp all day amidst the snowing and blowing. We were

METAMORPHOSIS

10,000 miles from home in a very wild country. It looked like snow again that night, and we had to decide to stay there that night or go down to a warm fire inside a primitive inn at the last village and drink beer. And we had to decide before dark.

This is where I learned a valuable lesson in leadership. Tension was high. Some men said they were going down, no matter what anyone else did. Some said, "Let's all go." With the wind howling, I had everyone stand up in a circle and speak up, one by one, to say what he felt — not what we should do or discuss, but "I feel..." "I feel scared, alone ..." The men wanted to say what we should do. I said, "Only say what you feel." Faster and faster. Around the circle. Cold. Exhilarated. Frightened. Lost. Ready.

We then voted with no discussion. One man and I wanted to stay. Everyone else wanted to go down the mountain. I knew that even if our tents blew down that night, we would be warm and dry wrapping the tents around us in our down bags. And that this was a spring storm that would blow out as fast as it would blow in. This was not a week-long winter storm. I also sensed the exhilaration the men would feel at waking up to clear skies with Mt. Everest punctuating the horizon. I made a decision to keep everyone together to do what the majority voted. I physically felt a sensation of letting go of my own personal desires to take the men where they wanted to go. No judgments, no letdown, no blame. A leader subordinating his personal desires to serve the team.

The sensation of satisfaction that I felt, knowing that I was serving the men, was deeper than being able to say "See, I told ya." Much deeper. So down the mountain we scooted, through the forest, snow blowing sideways, laughing and running. The monastery on the mountain side was almost unrecognizable under the snowfall. Across the river into the little village and the inn, smoke-blackened room around a wood stove, cold beer, hot chapatis, beds. And the requisite, shit-eating dog.

Thank you, men. I got to feel what it is like to be in service. Namaste.

If you would like to be on a mailing list for the next slide show of this trip or are interested in going on David's next trip, call (408) 479-1707 and leave your name, phone number and mailing address.

Rite of Passage

PETER ASHLOCK
Wailers ♂ EBNoM

BERKELEY

At the time I met Dr. Burden, I had not seen my father for over two years, and it would be another two or more years before we saw each other again. He was always in my heart, and I know now that some of my perceptions of him were the result of my own romantic nature mixed with truth, memory and wishful thinking.

Many of my happiest childhood memories were in his studio watching him paint, being shown how pigments were mixed, listening to him talk about great artists and the great movements in art. I loved the smell of turpentine and linseed oil. I loved watching him make things and improvise on what he didn't know. There were few places I was as happy as I was in his studio. Our casual conversations covered all manner of topics from artists to musicians to how things were made. We never talked about sports or cars.

In the early 1950's in Berkeley my father always managed to have a studio in some nearby storefront or, in one case, an unused bakery shed. I watched him draw and paint models he hired, watched him make stretcher bars, watched him stretch canvases and prime them. I modeled for him frequently (it kept me quiet and he didn't have to pay me). I learned how to sit still and hold a pose for long periods of time.

I can't remember when I first modeled for him. I only knew that my sister, who is four years older, had modeled also, and our modeling was taken for granted. My father would talk a little sometimes, and he would tell me if I could move and when I could take a break. Often we listened to music. Sometimes I was clothed and sometimes I was nude. We heated water on a pot-bellied stove and made green Japanese tea. It was idyllic in all the ways that the my parents' relationship with each other was not.

When my father left us and drove off to New York with all the paintings he could carry tied to the top of a borrowed Studebaker, I was unconscious of the enormity of my pain. But with the passing years came my increasing anger towards my mother, whom I blamed, rightly or wrongly, for his departure.

Several months after my dad left, my mother took my sister and me to live in Egypt, where she fulfilled a lifelong dream under the guise of a managing a public relations project for the Egyptian government. The cultures of ancient Egypt engendered a rich mythology in which the ideas of reincarnation figured prominently. This was a fascination of my mother's, and one that I accepted as easily as modeling for my father in the nude.

Living in Egypt was an adventure into a world both familiar in its humanity and its fabled history as well as exotic in its completely non-Christian, non-European orientation. It was the perfect antidote to the pain of my father's departure.

In Cairo I found out about the Islamic world, about poverty, and about a history so great and rich I have never absorbed it completely. In upper Egypt I saw magnificent temples and monuments; I had my first peak experience while visiting the temple of Karnak. I seldom thought about America and never missed much except my father, my grandmother, and our cat.

When we returned to our life back in Berkeley six months later, I began to look around and I liked less and less of what I saw in my peers. As the

METAMORPHOSIS

years passed and I did not see my dad again, I gradually began to invent myself in an image that I thought would reflect my father's values.

By 1962 the gossip regarding my nude modeling for my father had followed me for years and gotten nastier as time went by. I became less tolerant of the bullying by the testosterone-saturated little roosters who prided themselves on their athletic prowess and placed low value on any of the things I held as precious.

They seemed to live in a world where sex, cars and sports were the only relevant topics; my not sharing their values made me unpopular. That I was not embarrassed to have modeled nude must surely have disturbed them. What they could not know was that the implication that there was something dirty about this activity was an affront to my feelings about both my father and to the remnants of my innocence. At fourteen I was able to ignore the harassment only with increasingly gritted teeth, and one day I reached my limit.

I decided that whoever next insulted me and my father with these innuendos would receive my full fury.

In the lunchroom waiting in line for a tray one day, I heard a voice make a comment linking me, my father, and all that is filthy in one snide quip dripping with slime. I turned around and slugged the first boy I saw as hard as I could. I had never seen him before, I did not care who he was, I was angry and not going to hide it any more.

From that time on the comments seemed to cease.

I also began to refuse to participate in the gladiator-inspired team sports that passed for physical education. I didn't care who didn't want me on their team; I was determined to make them regret I was in the class. I was a bad sport and indifferent to the opinions of others. This was my only available revenge on the one area of their pride.

Although my father and I wrote and phoned each other regularly, this was but slight consolation against the intensity of my unconscious grief over his absence.

Maybe I was born angry, as my mother sometimes liked to say, or maybe I was angry about the dynamics in my family or the dynamics in the world. Whatever the cause, it is a fact that by the time I was fourteen I was a burning fuse waiting for an opportunity to express myself on my own terms.

In the summer of 1962 I began to effect a change in my public identity by refusing to wash my jeans and by purchasing a blue chambray work shirt. I wore turtle-necked sweaters and I avoided the barbershop. I saw as little of my mother as I could arrange. My career goals centered around being an actor, and I was in fact actively involved with a semi-professional little theater group. I fashioned my bohemian persona on what I imagined would dovetail with my father's world, and also on an article in Life Magazine which described a generic beatnik lifestyle.

A local public radio station offered a Friday night folk music program; I became a fan and began my education in the avant garde popular music of the period. This was an "open mike" show, and the station was usually filled with hangers-on and well-wishers of various sorts. It was the place to be cool, be seen and be invited to the inevitable party which took place afterwards, starting at one in the morning. I met others of like temperament — not many my own age, yet I was not alone. No one objected to my drinking wine and smoking cigarettes.

Early in the fall semester my mother heard of a man who was in town looking for students to take to his experimental school in southern Baja California. Another family had sent their son there for summer school and he had loved it. This was the era of Summerhill and many other alternative schools.

Eager to do something to change our relationship, my mother suggested I try this school. I'd have been willing to go anywhere with anyone, just to get away, especially to a foreign country. An appointment was made for me to have an interview. I met Doctor Burden and agreed to go to Shimber Beris, which was the name of this school.

Dr. David Burden was a tall Englishman with glasses and little hair on top of his head. I have to admit that the romantic in me was attracted to his being English and the adventurer was attracted to going to Mexico. I was promised formal education by correspondence course from the University of London. I was so impressed that I was going to completely outclass the jerks on the junior high playground that my imagination quickly took over; I was confident that this was the place for me.

My only regret in departing public school was leaving the drama class, where I was carving a niche for myself due to my theatrical personality and art background.

Ojai

It was arranged that I would meet the group in Fresno and that I must bring some specific equipment and clothing for living in the desert.

I was met at the Greyhound terminal by Dr. Burden, Dick from Berkeley and Rod from Long Beach. I wore motorcycle boots, blue jeans and a chambray work shirt.

The first thing we did as a group was to go to an enormous military surplus warehouse and purchase supplies for our life in Mexico. We bought large field stoves, pots and pans, tents, tarpaulins and two twenty-five gallon tanks for drinking water.

Dr. Burden had also just purchased a vehicle which had seen military service in North Africa in

WW2. It had bulletproof gas tanks, dual tires on the back, four-wheel drive and about a dozen various gears in two gear boxes. A game hunter had remodeled it with four bunks, a sink, a propane-powered refrigerator and stove. It was well suited for our purposes; we called it the "Elephant." That night we drove to Ojai, in Ventura County, to a place called Meher Mount.

Agnes Barron had been a foreign correspondent before the war but was now a devotee of Meher Baba, an Indian yogi. She had named her retreat on a mountain top overlooking the Ojai valley after her guru. There was a house, two dormitory-style buildings, a garage, a neglected swimming pool full of frigid algae-clouded water, and the overwhelming rotten-egg odor of sulphur everywhere. The area was known as Sulphur Mountain. Sulphur water flowed from all the taps and stained the sinks and plumbing.

The boys slept in one dorm and the girls in another. Regardless of whatever rules were laid down about what we could or could not talk about, all such rules were instantly ignored when Dr. Burden was out of earshot. I got a crash course in the life of a teenage surfer from Long Beach. Rod told me tales of surfing and sex and beach parties, woody station wagons and rock 'n' roll in the Southern California sun. I felt confirmed in my belief that he was from a foreign country.

The rest of our group included Dr. Burden's wife and their nine-year-old daughter, Daphne; two sisters whose parents were diplomats in Malaya; Tru, a tall woman with stringy grey hair who looked like a Dorothea Lange portrait from the dust bowl; Tru's husband Clear, who sometimes appeared in an old Kaiser car with a trunk full of produce rescued from some supermarket dumpster; Beth and Linda, another pair of sisters; and Brad, another boy from Berkeley, who was later joined by his mother Connie and little brother Steve.

METAMORPHOSIS

I attached myself to Dick from Berkeley; we had at least some common points of reference. He was four years older than me. Never having had a brother, I appreciated his tolerance and patience with me.

Dick was fascinated with working on Mrs. Burden's 1948 Studebaker and the 1933 Dodge coupe he found in Miss Baron's garage. I was more involved with letting everyone else know how unspeakably hip I was in all matters relating to the arts, controversial issues (of which at fourteen I had only an inkling), music and theater. Mrs. Burden tried briefly to teach us some theater techniques but, like much of the structure at Shimber Beris, this was lost in the shuffle.

Occasionally the group would go to the community center in Ojai to folk dance on a Friday night. And there were many different mundane activities such as getting non-sulphur drinking water in five gallon jars, picking oranges in you-pick-it orchards, cooking, doing laundry or washing dishes. I managed to carve some sandal soles from an old tire, using only a pocket knife and a lot of determination. On one trip to town I shoplifted a lovely Esterbrook fountain pen which made me feel very guilty every time I looked at it.

Most of the time it was the girls who handled the cooking chores and other traditionally domestic work. The boys were assigned to do labor-oriented work, such as fixing the septic tank or repairing the vehicles, none of which I knew anything about.

Every morning the group gathered in the living room for a meditation period. This could be silent, but more often it involved Doctor reading from some spiritual material of his choice: *The Third Eye*, by Tuesday Lobsang Rampa; *Winged Pharaoh*, by Joan Grant; *The White Company*, by Conan Doyle; or *The Moonstone*, by Wilkie Collins. And often Dr. Burden sermonized on whatever he was in the mood to critique; his themes generally revolved around his opinion that we were all the damaged products of a depraved society. The hidden agenda to be revealed was that only through Shimber Beris was salvation available to us, poor wretched souls that we were.

I learned that Shimber Beris meant "valley of the birds" in Swahili and that Dr. Burden was the son of missionary parents; he had spent a lot of time in Africa as a child. He was a doctor of chiropractic although he began to put on airs of being skilled in the areas of homeopathy and naturopathy and even dentistry.

After a few weeks at Meher Mount there was a falling out with Miss Baron, which resulted in our abrupt departure. For the next several months we drifted from one public campground to the next in Ventura and San Diego Counties. During this time I amused myself by getting away with smoking cigarettes on the sly, reading from the complete works of O. Henry and a copy of *Playboy* I had managed to smuggle along with me. The *Playboy* fed my adolescent libido and the short stories about turn of the century bohemia in Greenwich Village fueled my romantic notions of being a beatnik.

Often I accompanied Dr. Burden and Dick or Rod on seemingly interminable treks to find some piece of equipment our leader wanted to take to Baja: second-hand dental equipment, an ozone generator, a small grain mill, an arc welder, a washing machine, an electrical generator, tarpaulins, a fifty-gallon drum to hold gasoline, and a military surplus trailer. We searched health food stores for organic produce, raw milk, raw nuts, whole grains and honey-sweetened ice cream.

At a romantically seedy trailer park near San Diego I was caught smoking a cigarette in the shower by Dr. Burden, who told me that his wife had a psychic vision of me doing that. I forget the punishment; no doubt it involved some form of spiritual penance of his design.

The spiritual/metaphysical atmosphere unfolded around me day after day, although I was often amused that I had heard the references already from my mother in one form or another. The suburban children he picked up as students for his makeshift tough-love school of spiritual reintegration were often more confused, bewildered and submissive than I to the whims he proclaimed in the name of lofty spirituality.

Towards the end of my experience with the school I heard Dr. Burden casually claim to be capable of levitation and telekinesis, although he refused to demonstrate these skills in front of us.

My time unfolded in uneven pieces, now smooth and serene, now angry and rebellious. At times I felt sullen and resentful; at others I was happy to be Dr. Burden's pet clown; but always in the background were profound doubts about the purpose of my involvement. I reminded myself that the alternative was to return to the purgatory of the public school system, where I felt completely misplaced. I didn't know where I wanted to be unless it was at the High School of Music and Art I had heard of in New York City, and living with my dad. That was a dream never to be realized.

It seemed that new rules and new unpleasant surprises were always being unveiled. The rule that all mail home was to be read first by Doctor was a violation of privacy I was unprepared for. The idea was that he wanted to be sure we were telling the truth about what was happening to us. My level of communication with my mother at the time was poor enough that the content of my letters to her were bland and unremarkable in any case. The letters to my father, on the other hand, were much more direct and specific about my experience; those I simply smuggled past the censor.

BAJA

After five months of wandering around southern California in the Elephant, we were finally ready to go to Mexico. Our first stop was to be San Felipe, where a boat the school owned was supposedly docked.

In Tijuana we stopped only long enough to find drivers to guide our two-vehicle caravan down the peninsula. Just past San Felipe we made camp by the side of the road in the desert less than a mile from the beach. This turned into a several day stopover. It was discovered that we lacked some crucial automotive parts for the Studebaker. Dick and Rod, being the oldest, were sent to get replacement parts.

That night we were told that the school's boat had sunk. The story was that a storm had torn it from its anchor and demolished it totally. We walked the beach wondering which pieces of driftwood had once been part of the boat. Later a rumor spread that the Mexican government had taken the boat due to some disagreement with Dr. Burden.

While we were camped waiting for Dick and Rod to return, I practiced baking bread in the gasoline-fired field oven and generally enjoyed the warm spring weather. The totally exotic wilderness and the absurd circumstances of my existence suited my sense of irony.

One American tourist stopped with his camper to chat briefly, unaware of the nature of our group. "Hyuck-hyuck, we hadda catastrophe last night," he chortled. "We run outta liquor." Another American tourist drove past towing a boat called *The Wet Dream* ... this was the America Dr. Burden loved to hate and he seldom wasted an opportunity to remind us that we were all products of the depraved society north of the border. What I loved was watching his reaction to these tourists, which struck me as equally absurd as the vulgarity which offended him.

METAMORPHOSIS

At last we continued our trek south. The changing desert delighted me. The landscape with its austere and unexpected beauty was overwhelming, and I was fascinated with the occasional villages along the road.

In March the full force of Spring was on the desert — balmy weather, blue sky, jackrabbits, birds and exotic vegetation. We stopped in each village to stretch our legs and to buy *penuche*, the crude sugar sold in every general store and which in every village tasted remarkably different. Some tasted like light molasses, some more like un-sweetened chocolate, some almost like maple sugar. The general stores were like sets from a western movie, with all the goods one would need sold from behind one long counter that ran the length of the store.

We managed to average a hundred miles a day and never failed to have at least one flat tire. At night we camped at the side of the road, built a fire and cooked our vegetarian meals on the gas burner in the Elephant.

Frequently the meals consisted of *atole*, a cornmeal mush eaten by the people in the region. Postum, a roasted grain product which supposedly re-sembled coffee, was the hot beverage of choice, sweetened with brown sugar.

The native navigator/drivers always made their camp at a discreet distance from our group. I watched as they built a simple snare and later roasted a small animal over a fire of their own. I admired them, and part of me wished I could learn from them what their lives were like, and not live within the fantasy culture created by Dr. Burden.

As we crept day by day down the peninsula there were always new images to contemplate, such as the Viscaino desert and the waters of Scammons Lagoon, where whales came to give birth. On the broad sands of the lagoon were scattered beautiful turtle shells; I didn't know it then, but the flesh of these turtles was the plunder of hunters who

would later themselves be the target of environ-mentalists.

Somewhere along the road we stopped for a day so that the older boys could change the transmission in the Studebaker. At another juncture we bided our time while they replaced an axle on the Elephant, which, with all its power and strength, had met its match in this terrain. One afternoon we came upon some Americans whose jeep had cracked its frame crossing a ditch. They were more than surprised when we pulled over and offered to weld the break for them on the spot. It was fun to be a part of the group at these times.

At other times I could stare at the beauty of the landscape and tune out the surreal contradictions I felt bombarded and threatened with from Dr. Burden and occasionally from the other students. Dick had quietly counseled me early on to "take what I needed and leave the rest." His words helped, but he had demons of his own and was prone to be distant at times. In time I would find it harder to "take what I needed" as my anger at new insanities blinded me to what goodness there was.

One afternoon after more than ten days of deeply rutted and rock-strewn roads, we came at last to "good" road which had been scraped flat by a grader. Dr. Burden told us that we would soon be in La Paz and near the end of our journey. For me the best and worst was yet to come.

Some hours later we pulled over to the side of the road near a deserted stretch of the beach on the bay of La Paz. Dr. Burden found a telephone or telegraph facility and contacted his hired helper in San Bartolo, Juan Alvarez, who drove up in another vehicle owned by the school, a Dodge Power Wagon once used as a military ambulance in WW2. Juan was to become my most trusted friend for the rest of my stay with Shimber Beris.

Our caravan drove south on the main road from La Paz toward Cabo San Lucas. The road was sandy and rutted in places, passing through shallow arroyos and rocky outcroppings covered with enormous wild fig trees. Cactus in vast varieties along with a plethora of other desert vegetation seemed to grow everywhere. We passed through quiet villages, a ghost town and the first vestiges of what would later become "resorts." Eventually we arrived in San Bartolo, about halfway between La Paz and the cape.

The excitement of actually living in a foreign culture overcame the doubts I had about being with the group; for the moment I was more than satisfied to have come this far.

San Bartolo is an inland village situated on a large spring which issues forth from a low cliff in the middle of the village. The village was occupied by people of modest education and economics, mostly involved in agriculture. In every way they appeared to me to possess a sweetness and dignity that is missing in American city-dwellers. The women wore long print dresses, and most had long black hair; their feet were either bare or sandaled. The men wore denim pants, blue or white shirts, straw hats, and sandals made from rubber tires and inner tubes.

Dr. Burden had leased a piece of land from a Señor Castro, whom he had dealt with before. The land was about an acre in size and had one building on it. An irrigation ditch ran along the lower edge of the property and separated it from a lush orchard on the other side. At one edge of the property was a large rocky outcropping with the ubiquitous fig trees growing on it. The outhouse stood to the far north. There were small palm trees scattered in the field, which lay well below the main road. The driveway into the property was steep and badly rutted.

The building, next to which we parked the Elephant, was built of stone; it had one window and a sturdy door. From the south wall a very large roof slanted broadly down to a point some distance from the building; the three remaining

sides were open. The area under the roof was packed dirt.

Beside one of the stone walls a traditional cooking fireplace had been built. This was an elevated hearth on top of which were constructed three trough-like spaces divided by mud bricks. A kettle placed over the cooking fire would rest on the edges of the two parallel rows of bricks. Most of our cooking was done here; the remainder was done on the field stoves.

The area under the roof was the main social area for the group. A large dining table was installed — and soon there were some unusual rules in place as well.

We were forbidden to speak under the roof at any time. And we were forbidden to speak about the past anywhere under any circumstances. The punishment for breaking either rule was to be forbidden from speaking for an entire day. The punishment for violating one's punishment added a day to the sentence. It looked for a while as though some of us might never be allowed to speak again.

Our tents were set up in the field with generous distance between them. I was not eager to sleep with my face at ground level, the ground being very dusty and hard, nor was I eager to be at eye level with scorpions, snakes or iguanas. I constructed a bed by first cutting four limbs with a "vee" shape at one end and sinking them into the ground for corner posts; I braced them with rocks, sticks and mud. I cut two limbs for the width and two more for the length; I laid these pieces in the crotches of the "vees" and lashed the frame together. I wove the springs for the bed out of *ule*, which is made by cutting a continuous strip from an inner tube, usually for the purpose of tying down a load on a truck. I had the best bed in camp.

Various field trips became routine. The group would pile into the Dodge Power Wagon to go get

METAMORPHOSIS

the trophy fish the rich Texans often discarded after having their pictures taken, before they flew home in their private planes; we ate a lot of free fish this way. Other trips were to small farms to procure vegetables or fruit.

℘

Slowly there seemed to grow upon our group a vague nightmarish feeling of oppression and a cold manipulative hostility from Dr. Burden. Mrs. Burden had been installed in a villa in La Paz, and we missed her gentler influence. Occasionally I would see her in La Paz when we drove up to buy things not available in San Bartolo. Often the students talked about the feeling that we were living in a concentration camp. Frequently many of us were depressed.

On one visit to La Paz, Dick and I sat on a bench waiting for Dr. Burden when we saw the actor John Wayne come ashore from his converted minesweeper, anchored out in the harbor. I remember being amused and mildly repelled by these Hollywood Americans, who seemed more alien than ever to the world I now occupied.

Another wonderful memory of La Paz was the opportunity Dr. Burden found for me to spend a day in a bakery watching the traditional methods of baking with a huge brick oven being applied by a generous and friendly crew of young men.

These became the moments of grace between the extended passages of emotional turbulence fueled by the sometimes sadistic emotional treatment from our guardian with all his spiritual rationalizations, all his threats of "bad karma" or the horrors that life after death would bring if we didn't straighten out now.

At times like these I retreated within myself, into a shell of protective memories where I nurtured all the details I loved about my home in Berkeley. I remembered the trees, our cat, the house, the rooms in the house, special Sunday mornings

when we listened to music and my mother and sister played Scrabble. More than anything else I returned to the memories of music which I longed to hear again. As a child I had played violin and developed a strong ear for musical memory. This gift kept me as close to sanity as I was likely to come. I remembered favorite passages of Mozart and Hayden, of Broadway musicals, and of the new Joan Baez records I had listened to avidly not long before leaving Berkeley. I read short stories in the O. Henry collection, looked at pictures and read from *Playboy*.

Somehow or other my father sent me a copy of The Realist from New York, an outrageously hip underground satire magazine which, had it ever been found, would have guaranteed me a place of honor in the special hell reserved for those who displeased Dr. Burden.

I also amused myself by carving wooden spoons from wood I found in the desert on a quest for better materials for my bed.

I was deprived of the pleasure of my bed on my birthday, however, due to an incident which began to shift the balance of my feelings against remaining at Shimber Beris.

Repeated disagreements with Dr. Burden, we were warned, would lead to punishment more severe than a mere "silence" for a day or two. The offender would be put into "isolation" — cut off from all communication and interaction with the group — and be deprived of all familiar or protective environments.

I have long since forgotten what transgression brought the wrath of Doctor Burden upon me but the ultimate in his arsenal was laid to my psyche on April 16, 1963, the day before my fifteenth birthday.

I was confined to one square yard of ground next to a palm tree, off in the sticks, well away from the camp but in plain sight. I was required to stand, not lean against the tree, with my back to the group. A tray of food was brought to my feet three times a day. At night I was allowed to sleep on the ground where I had stood all day.

Finally after three days it was decided I needed no more of this therapy, and I was elevated to a bottom-of-the-pile kitchen helper. I cleaned all the dishes for the school after each meal. I was still being given the "silent" treatment, but I was allowed the privilege of sleeping in my tent again. This lasted another two weeks.

I no longer recall the process by which I re-emerged into the better graces of the headmaster, but I learned that I might as well lie with a smile on my face if that relieved the abuse. The seeds had been planted for my own insurrection.

Mine was not the only rebellion. Two boys ran off to La Paz but were picked up by the police and

ORIGINAL PEN & INK DRAWING BY
THE AUTHOR

returned. I slowly formulated a different kind of rebellion. In the meantime, I was determined to glean as much pleasure from the beauty and charm of the land and the culture as I could.

Our most interesting excursion by far lasted several days in the coastal desert, where we helped build the first road from the main highway to a village called Cardonal.

Don David, as the natives called Dr. Burden, rode a horse, as did his daughter Daphne, making some of us feel like peons with our great white master while we labored with picks and shovels in the hot bright sun alongside a crew of local people who would benefit from our efforts.

We drove the first automobile into the village. They told us that many of the women and children had never seen a real motor vehicle before, except in magazine pictures. I felt ambivalent about our distinction.

Cardonal is a hamlet of thatched houses on the Sea of Cortez. Fishing is the principal source of income. The men sell their catch in La Paz, transporting it in their canoes. In this village I watched two men spin horse hair into a beautiful bridle for a horse; I was fascinated by the process of the craft.

In the desert we gorged on fresh cactus fruit as red and sweet as watermelon, getting our hands bloody from the thorns. We camped on the beach for several nights, and one of the most spiritual experiences was watching the sun rise over the Sea of Cortez. Yet there was real danger in this environment. We swam often in the clear water, and many of our group were stung by the ubiquitous man-o-war jellyfish with their poisonous tails.

Another kind of poisoning came from the free fish we were given one day at Los Barriles, a resort; the school devoured this fish greedily for dinner. I awoke in the middle of the night with the most intense itching of the palms of my hands and the

METAMORPHOSIS

soles of my feet. The next day I found that others had more violent reactions to the food poisoning.

More than once we ate rancid butter. Dr. Burden bought weevil-infested wheat which we cleaned tediously by hand before grinding it in our little electric mill which was powered by the enormous marine generator we had bought in San Pedro.

Our general sanitation was probably no better than anyone else's in the village, despite our electrical generator, which allowed us to run the washing machine, a light bulb, and whatever Dr. Burden had in the office he'd set up in the Elephant.

I bathed usually in the irrigation ditch at the edge of the property, often stealing mangoes or papayas from the orchard next door where I loved to wander.

As the summer progressed I began to notice I had sores on my feet that did not heal. Feeling myself to be a hardy sort, I ignored this condition until I began to realize that the sores were spreading; in fact, if I merely scratched uninfected skin with my fingernail, I would have within 24 hours a running sore at that place, festering and spreading like the others. My legs hurt from the inside and were only comfortable when lifted horizontally. I was told I had subclinical pellagra, related in some way to leprosy. I saw people in the village hobbling around with legs covered with the sores and began to wonder what this all added up to for me.

When I could, I spent more and more time talking to Juan, confiding in him my anger and frustration towards Dr. Burden. He was entirely sympathetic and told me that Dr. Burden had changed, he was not like he used to be, and it disturbed Juan too.

On one occasion I visited Juan's house, a woven basket-like house with a thatched roof, off the main road close to our property. His wife was cooking, and I commented on the delicious smells.

Soon after that Juan began to regularly bring me food his wife had cooked, a delicious memory I still savor as much for the sensual pleasure as for the incredible generosity of spirit.

Before the problems began with my legs one of my main pleasures had been to walk through the village and buy eggs from people who had extras to sell. This gave me further opportunities to witness their lives up close in ways which I appreciated more than I could adequately express.

A visit to the butcher was a revelation in the matter-of-fact realities of the operation in such a primitive situation. Under a half roof lay the carcass of the cow where it had fallen in death. The butcher cut off the pieces fresh from the carcass for customers on request. When business was slow, he cut strips and hung them on racks to dry in the sun.

The reduction of mobility from my diseased legs meant that the only peace I had was when the group went off on a trip and I declined to join. I could be alone or visit with Juan, who came to help me speak Spanish. But I was getting to the end of my patience with this experience, and I was beginning to formulate the plan for my final rebellion. I regretted that I'd be leaving Juan and everything else that I loved there. I loved the sounds and smells of the village, the donkeys braying in the night and the roosters crowing before dawn. I loved Matilda, who spoke no English and had come to cook for us and help with laundry. She ironed the clothes with old-fashioned irons heated over the coals; to dampen them she took a mouthful of water and liberally sprayed the garment as she held it at arms length. She made flawless tortillas with her hands as fast as a machine. She was graceful and patient and kind. She and Juan were no doubt among the better-paid members of the village. I would miss them both.

But the emotional claustrophobia of the situation, my deteriorating physical health, and the fact that I wanted to see my sister (who was leaving Berkeley soon to go to school in Italy for a year), all prompted my decision to leave.

Because of my awkward relationship with my mother, she had no idea how frightening and desperate my situation was. Only in the letters to my father that I routinely snuck by the censor had I described my experiences truthfully. The communication between my parents was limited and poor. The result of all this was that my mother disbelieved my complaints about the school when I finally described them.

I decided that my best strategy was to force Dr. Bruden's hand by randomly and routinely breaking all the rules. This meant talking openly with girls about sex, always a popular topic at fifteen. It also meant talking openly about the past, and talking under the big roof, and continuing to do so with a smile on my face even after I was punished with the assigned "silences." I forget how long this went on, but I know I managed to aggravate Dr. Burden to the extent that he saw me as a threat to his control. He wrote my mother and told her I would have to leave. Once again I was put in "isolation," this time in my tent because of my diseased legs. By this point I was glad to be alone.

My mother wrote an angry letter to me and told me I would not be allowed to return home but would go straight to a military boarding school in the suburbs. I figured I was ready for a change anyway, regardless of what direction.

I had a hunch that Dr. Burden would not easily let go of the two thousand dollars a year he was getting for me. A few minutes before I was to leave with Juan to go to La Paz, he called me into his office and told me he was sure that if I wanted to stay that we could work things out satisfactorily. I told him I was quite sure that I was far too disturbed for him to be able to help me.

I felt very melancholy about going back to the States. There was so much that I was ambivalent about. I was sad to leave Juan and the village and the wonderful landscape of southern Baja California. I hugged Juan in tears and waved good-bye and boarded the plane.

Years later I met someone who knew someone else who had gone to Shimber Beris after me, and they said Juan had died of diabetes a few years later. Tears pricked my eyes.

On the plane flying north to uncertainty I picked up a news magazine to pass the time and saw a picture of a holy man in a saffron robe in flames in a country called Viet Nam. I didn't know where that was and was completely shocked by the image of a burning monk.

Epilogue

Shimber Beris touched my life only a few more times after that, but always unexpectedly.

When I returned home I was not sent to military school but, much to my complete bewilderment, was brought back home as if nothing disgraceful had occurred. It was discovered that my skin disease could be cured quickly with a balanced diet and good sanitation, plus a little antibacterial ointment. The scars lasted for years.

My mother was slow to acknowledge the truth of my experience and it wasn't until I was visited by other ex-students and we discussed the events in front of her that she realized I had not made it up. My father was glad it was passed, but he had been furious at the time.

Dr. Burden came up to Berkeley some months later and bought an ocean-going tug boat which he berthed in the Berkeley marina for a time. He had dinner at our house at least once and I visited them at least once. He wanted me to call him

"Dad" but I was not able to bring myself to quite that level of affection for him. He gave us a beautiful Siamese cat, named "Popocateptl," which I adored until he ran away.

Shimber Beris faded into the repertoire of stories from my life until the fall of 1986, when I was a cab driver in San Francisco. I was in the midst of a collapsed marriage and in great emotional turmoil. I picked up a Sunday newspaper and saw Dr. Burden's picture on the front page. The headlines said something about a vision of utopia that had died in the jungle in Guatemala. I read, with tears running down my face, about the death of the school, the abandonment of Dr. Burden by his wife and daughter. I read of students who had remained there, married to other students, now with children of their own; they had been there when I was there almost 25 years earlier. The same sort of chills ran up my spine as when the People's Temple suicide took place. I was glad I had known when to leave.

ORIGINAL MASK SCULPTURE BY THE AUTHOR

A Dispatch from the Very Edge of the Known World

BILL BURTON

♂ *SBNoM*

MARCH 1993

The following is a dispatch from the Burton Expedition on its mission to the outer reaches of the known universe, Raleigh, North Carolina. It is intended to be a travel log, an encouragement, a warning and simply an update to my friends about how the expedition is going. I know some of you are planning such expeditions of your own and I hope that some of what is contained in this dispatch will either encourage you or panic you as the case may be.

This expedition is just a little over three months old so we are still a little amazed at the new sights and sounds of this far off, strange and yet wonderful new world. We have met some of the natives and a few missionaries from other fabled lands like Chicago, New York and New Jersey. We have even met a few travelers from California although they didn't want to admit being from the "Left Coast" as it is referred to here.

We have learned to understand some of the strange language that is spoken in these here parts, although we have yet to begin speaking much of this unfamiliar tongue. The one word which has seemed to come pretty naturally, sort of like "Aloha" in Hawaii, is "y'all." This is a very functional word and I think we should consider adding it to the American language. We might have to work a little to get New Yorkers and those people in Boston to accept it and pronounce it correctly, but I have faith that they are trainable with a little effort.

Please note that this word is "y'all" not "U-All." A subtle and yet important distinction lest ye be known in these parts as a Yankee. Sort of like calling San Francisco, "Frisco."

DISCLAIMER

I don't mean to belittle or demean San Jose or California or the people who live there. I'm just trying to draw distinctions based on the common experience we share which is living and working in San Jose. I didn't like it, and I still don't. I have NO interest in moving back and I would encourage anyone who dislikes it as much as I did to get out. For those of you who like it or love it all the better for you. If you can get rid of the people like me who don't share your passion you can have things back the way they were when there were fewer of us and more of you. It was a beautiful time and maybe it will come back once we all get the hell out of the way. May California again glitter with all of the gold which it so richly claims to have in some bank in Beverly Hills in somebody else's name.

Before we begin, two stories to set the tone and context for this dispatch:

SEEING THE ELEPHANT

There is an old story about a man who one day reads in the newspaper that the circus is coming to town. He gets very excited because every circus has elephants and while he has heard a great deal about elephants he has never actually seen one with his own eyes.

So on the day the circus is scheduled to come to town he loads up his cart with all of his watermelon crop which he will sell at the market, and heads off toward town to see the elephants. As he enters the town square, there to his surprise is the circus just arriving and parading down Main Street. So the farmer stops his mule drawn cart and joins the growing crowd to watch in hopes of seeing his first real live elephant.

Well sure enough, there in the third or fourth row of the parade is a huge creature which he immediately knows by its size and color must be an elephant. It's much bigger than he ever imagined and as it draws near he can see the large ears and the trunk, and oh, those massive feet.

Now, the farmers mule has never seen an elephant either. It is a lot less impressed with the sight than the farmer. In fact it is damn right scared to death of this ... this ... thing that is coming straight at it. The mule bolts and in so doing turns over the cart and spills all of the watermelons on the street where they are broken into a million pieces and then trampled by the elephant.

The farmer looks down and realizes that his entire crop is lost. And in that moment realizes that while he got to see the elephant it has cost him very dearly to do so.

I have to say that this story sums up better than I could my experience in California. I got to see the elephant but boy did it have a cost.

PIG!

So one day here in North Carolina I'm driving down this dirt road and, as I sometimes do, I'm driving a little faster than I probably should. But there is no traffic and it's way out in the country so there are probably no cops within miles, so I'm pushing it a little just for fun. As I round this bend, I look up and here comes this pick-up truck from the other direction but in my lane.

Well, we both hit the brakes and I start spinning and the truck starts spinning and dust and gravel is going everywhere. I'm still not sure how but we manage to miss hitting each other and finally come to a stop about 10 yards apart. As the dust begins to clear I look back and here is this old woman getting out of her rickety old truck. She looks at me and yells "PIG." Well, I'm not one to take that from some old country hag so I roll down my window and yell in my best man voice "You old COW!!!" Then I gun my engine and head off down the road throwing a ton of dust in the air.

As I round the curve I look up to see a PIG crossing the road. What that old hag was doing was warning me about the pig. It just goes to show you have to watch out for your preconceived ideas about people, places and things.

GEOGRAPHY 101

Raleigh is in North Carolina, not Virginia or South Carolina. It is at Latitude 35° 47'N and Longitude 78° 39'W, placing it 1½ degrees south of San Jose, California (37° 20'N, 121° 53'W). Raleigh is the capitol of the state of North Carolina. It is about

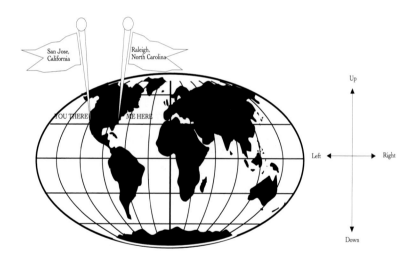

80 miles west of the Atlantic Ocean, about 200 miles east of the Great Smoky Mountains and about 60 miles from the Virginia border to the north. The area is primarily gentle rolling hills forested with pines and hardwoods like oak, maple and dogwood.

We live in a development that is 16 miles north of downtown and 2 miles south of 33,000 acre (read

big) Falls Lake. It is 10 miles to the famous (?) Research Triangle Park, the Silicon Valley of the Southeast. Falls Lake is formed by a flood control dam across the Neuse (pronounced noose as in hangman's) River. From below the dam one can river raft or canoe to the ocean, a trip I hope to make this summer. The lake provides excellent fishing and water skiing.

If you want to get a flavor of the land, make a point of seeing two recent movies. Both *Last of the Mohicans* and *Prince of Tides* were filmed almost entirely in North Carolina. *Mohicans* gives a great visual feel to the mountains of western N.C. and *Prince* a good feel for the eastern outer banks area.

HISTORY PART 1 — POLITICAL

Named for Sir Walter Raleigh, this is one of the original 13 colonies and so was granted statehood in 1776. It was an important state in both the Revolutionary War and the Civil War. In the Revolutionary War (you know, fuck the British and all that) the first decisive victory was won by the Colonists just a few miles from my house. The final surrender of the South to the Union occurred here in Raleigh just a few days after President Lincoln's death. No, it didn't happen at Appomattox, that was the surrender of one army

to another (Lee to Grant). The final surrender of the two sides took place seventeen days after Appomattox at the Bennett Farm House, 8 miles west of Raleigh, when Johnston surrendered to Sherman.

Don't let the "North" in North Carolina fool ya' — it was on the side of the Confederate States (The South). But one thing we are learning is we didn't learn much about the Civil War in school. What we did learn was greatly simplified, written by the winners and/or just plain wrong. It was a very interesting time and to be living in the heart of the country where it happened places a whole different perspective on what happened and why.

One of the reasons we chose North Carolina is its proximity to the early history of the U.S. I never liked history much in school, but a couple of years ago I read a book titled "Don't Know Much About History." It was sort of a history book for those of us who didn't get it in high school. I came away fascinated by all I didn't learn when I **had** to. I'm looking forward to the spring and summer when we can make some weekend trips to the many historic places within a day's drive. Of special interest are Washington, D.C. and the Smithsonian Museum.

HISTORY PART 2 — TOBACCO

A young man returned from the Civil War to find that his farm, including his home, crops and personal possessions, had all been pillaged. His only belonging was a sack of tobacco which he had hidden before he left for the war. He and his friends used sticks to beat the tobacco on the floor of a corn shed until it was pulverized enough to be usable as smoking tobacco. Then he sold it to earn the fare to travel to New York where he started a small tobacco business. And then in 1888 bought the exclusive rights to a new machine that would roll tobacco into pre-fabricated cigarettes. His company, American Tobacco went on to be the largest tobacco company in the world. That was

until 1912 when the government declared it a monopoly and broke it up into four smaller companies: American Tobacco, Liggett Myers, R.J. Reynolds, and P. Lorillard. Kind of makes you wonder what would have happened if his farm hadn't been sacked during the war. He probably would have been just another hayseed farmer. Instead he became one of the best known philanthropists of the time — his name: Mr. Washington Duke (as in Duke University).

The history of this area as a tobacco growing agricultural community is evident in the number of small farms scattered around. Each one includes one or more tobacco drying sheds that are very strange looking "barns." It took me several weeks to figure out why the barns here were so tall and

narrow. These sheds, especially the older ones, are very interesting in design. They are constructed on stone foundations and are planked with quarter sawn timbers which are caulked with a combination of cement and mud. Inside the farmer would build and stoke a small fire which dried the tobacco which was hung from the tall roof structure.

The whole idea of tobacco growing is as mysterious to me as the idea of fruit drying sheds are to newcomers in San Jose. I'm sure they are plagued with the same stupid questions, like why would they dry fruit in the middle of a city which makes silicon wafers? They don't remember and there is little evidence of the farms, orchards and ranches that just 20 years ago supplied the economic force behind what was one of the most prolific agricultural areas of California. The same thing here. Much of the infrastructure that supported an agro economy is giving way to the new technology-based industry. What is here are a few confusing clues about what used to be. Like in downtown Durham (15 miles west) the entire town is surrounded by **huge** brick buildings in various states of disrepair. Now abandoned, they give some indication of the volume of tobacco that was grown, processed and turned into smokes by the Liggett Meyers Co. And that was just one of the four largest tobacco companies in the world.

The Job Market

Both Diane and I have been looking for work for just over two months now and while it is slower than either of us would like it is probably going about as well as one can reasonably expect. We moved here not knowing anyone in the state much less the town, so much of our efforts to date have been in getting a handle on where to go, who to talk to and such. We have also spent a good deal of time and effort building a network. Just getting to know people and having them get to know us. I have now contacted by mail, phone or in person 242 people. I have joined some organizations for the purpose of meeting people and as much as I hate this stuff I have been pretty good about

getting out there. This is not one of my strengths and it takes a lot of work for me to stretch like this.

People I — The Demographics

People here fall into three categories. The first is those who are from here ... natives. Generally referred to as "Inside-the-beltline" folks. They are traditional Southerners. No I don't mean Gomer Pile, Andy Griffith and the rest of the gang from Mayberry. I mean very conservative, very genteel, very ... southern. They are still the majority (I think) but are rapidly being outnumbered by the second group, the "new guys" from everywhere USA. For the past ten years this area has been growing by leaps and bounds (approximately doubled in the last 7 years) due to companies moving large numbers of workers and executives in from all over the country. It is very common to find most people at any gathering to be from somewhere else, typically New Jersey (IBM, DuPont, and BASF), Texas and Boston. The vast majority of these people **love** it here and are not interested in moving anywhere else including back from whence they came. Then there is the third group, into which my daughter falls, those who are here under extreme duress and will escape at the first opportunity. They hate it, loathe it, think it's Provincial, Dull, Ignorant, Slow, Unsophisticated, and Backwards. These people are in a definite minority and most are from New Jersey. I guess once you've smelled the garbage from New York no place else is ever going to be good enough again.

To break these groups into percentages:

Group I	Inside-the-Beltliners	55%
Group II	Ain't-Goin'-Nowhere-Else	40%
Group III	When's-The-Next-Bus-Leave-for-Jersey	5%

As I see it, this means that 95% of the population loves it here and is in no hurry to go anywhere else.

METAMORPHOSIS

Compare this with San Jose where I figure the breakdown is something like this:

Group I   Natives   5%

Group II   Love it and wouldn't live anywhere else   5%

Group III   Just working here so I can go somewhere   90% (anywhere) else

So, my point is that it is great to be somewhere where people want to live there. And the few New Jerseyites who don't ... well let's consider the source.

PEOPLE II (A LITTLE LOCAL COLOR)

"Hi, I'm ya nayba fum ova thar. My name's Boz, thaz showt fo Bozwuth (Bozworth)." That how I met our neighbor from back behind our property. He doesn't live in the development that we live in but his property backs up to ours. It's probably 100 yards or so through the wood from our house to his. In the summer there is enough foliage on the trees that you cannot see his place from ours, or I guess ours from his. But in the winter, once the leaves fall, his house is just visible through the woods.

Well, Boz is about 70 and as he tells it he and his family have lived on this land for 15 generations. His Grandmother was born on the land the year the Civil War was over. As a kid he ran, played and did boy stuff where all of these new houses stand today. His family sold off most of their 140 acres to the developers but as far as he is concerned it's still his woods. He lives in a smallish rough build house with a tin roof on the last 26 acres that they still own of their original homestead.

There is a small creekette that flows through our backyard and then through Boz's property. Then on into Barton Creek which then feeds into the Neuse River at Falls Lake. He tells me the other day that this little creekette as I've called it is known to the locals as Still Branch of Barton Creek. The reason they call it Still Branch is that just above where our house is today on the banks of this little creekette was where his Granddaddy had his moonshine still when Boz was a young man (yung-un).

LITTLE KNOWN FACTS

Did you know that Babe Ruth hit his first major league home run in Fayetteville, N.C., or that the Civil War ended here?

The forward pass which revolutionized football is a local creation as well. It was "invented" as an illegal play in a game between University of N.C. and Georgia Tech in 1895.

PEOPLE III — BUSINESS MEETINGS

All business meetings start with 10 to 20 minutes of social talk. You can talk sports, cars, guns, toys, weather, or anything else. Just don't talk about business. After you have sort of settled in and gotten to know a little about the other person you can start on the business topic. This part of the meeting goes pretty much the same as in San Jose. The people I have met are smart, well-educated and interesting. No Gomer Pyles here. Then, when the business is done you need to switch back to

"not a business" topic for 5 to 10 minutes to complete the transaction. This means that meetings here last an average of 1½ hours compared with 1 hour in California. At first this social or relationship stuff seems to get in the way of getting stuff done but in two months I've come to sort of like or at least appreciate the clearing it provides for getting down to the subject at hand. You just have to leave time for it. It is very impolite to rush through this portion of the meeting so a little patience goes a long way.

The Relationship with Diane

I think this has been a real growth experience for my relationship with Diane. We have had to count on each other and work together in ways we never had to before. It has brought us closer together and made me appreciate her even more than before for her power, softness and independence. She has made so much of the drudge work seem easy by just breaking it into small gorilla sized chunks. We have had some fun, some fights and some great sex. I think the release of the pressure cooker that we lived in San Jose has helped us both have more tolerance, patience and time for each other. We probably have more time together than either of us would like but it's OK and we both find time to do the things we enjoy apart.

Eastern Chapter of the Nation of Men

I have met several men here but none that I feel close to yet. I have met the brother of an old friend of Coogan's here. He is planning to do the next Men's Weekend in Boston and there is one other man who has asked questions about the Weekend and is probably a good candidate in time. Diane and I are thinking of running an ad in the personal section of the local newspaper to see if we can find men and/or women who have done the Weekend. With so many people moving here from Boston and California it seems likely that there must be a few who have done the Weekend.

In the meantime I am staying in touch with my team and keeping up on what's happening with the Nation of Men. The truth is, it is very hard to establish the level of relationship we all take for granted. I have days when I miss the men. Especially on days when things go very good or very bad. These are the times that in San Jose, I would reach out and have lunch with one of the men or have some port and a cigar with a man. These are difficult times to recreate without the context of the Nation.

What I really find missing is an outlet for my passion. In San Jose I had Microbar, my team and the Nation where I could express and stretch my passion. I had a big vision and men around me who would follow where I led and who acknowledged me even when I failed or fucked up. That environment gave me the courage to try lots of new things and ideas. I knew that the men would not let me get too far out of bounds without someone yanking my chain.

I was also in a continual learning process about leadership, friendship and what it meant to be the man I always wanted to be. That's tough stuff to give up and I miss it! The other thing I miss is being in an environment where I had paid enough dues that I was accepted as a man who could be trusted and who, even if I was off track, meant well. No one here has that experience of me, so I enter each new interaction with a score of zero.

I also have to say that I simply miss certain men. Just having them in my life once a week, once a month or every once in a while. I knew they were there if and when I wanted them or needed them. I won't make a list here because being the jerk that I am, I would probably leave off one or two of the most important ones. But, as I scanned the Resource Directory I kept seeing names and faces that had shared and given so much over the past 8 years. A man who held my hand and helped me climb a hill I never thought I could. A man who showed me what was on the other side of my anger if I just did the work to get there. A man who trusted me with his life. A man in whom I entrusted my life. A man who is the most creative mind I have ever known. A man whose spirit is with me every time I see the simple gift he gave me. A man with whom I rowed the canoe ... downstream. A man who beat me till we both bled. A man who had the courage to share enormous grief with me. A man who pisses me off because he reminds me of me. A man who is good enough just as he is, but doesn't know it yet. A man who has the courage to confront every barrier in his life ... but one. A man who, learning of my passion for seeing Montana, gave me a map and a kick in the ass.

ADVICE

Take advantage of the support structure that exists in your teams and in the NoM. Take chances, do new stuff that's hard. Take on new challenges and go for big risks. You may not always have that safety net to fall back on so make good use of it while it is available to you. Trust the men they will tell you if you get too far out on a limb or are in danger of hurting yourself or someone else. I think most of us are playing too small and too safe. Have Cross explain the theory of "Scream and Leap" and then go find a really big Kzin and kick it right in the nuts.

Drop all of the story about "not enough time," "too busy," "I would but ...," and stretch big while you have these men to give you support and a safe landing. You can jump from the plane without a parachute. The men in NOM will never let you hit the ground. Take it from a man who has made many, many, many such jumps and who has never hit the ground. Except of course for the time when I actually did jump out of a real plane **with** a chute and broke my leg. But that's another story.

BOTTOM LINE

I do not miss the smog, the traffic, the foreigners who don't speak engrish, the noise, the congestion, the freeways, the high cost of everything, not owning a home, the nightly news of death, destruction and mayhem, the police helicopters at 3 a.m., police sirens in the night or the drugs in my daughter's school. I Love It Here! It is all I hoped it would be and more.

I miss my son, the men, Mexican food, Cosentinos, Donna at Mission Pipe, Nation of Men, Team meetings, the Delta, Chevy's, Avocados, fresh vegetables even in winter, getting dozens of phone calls a day, and being counted upon by the men.

I hope you have found this dispatch interesting, fun and/or thought provoking. If not, why did you read it?

BONUS

BURTON'S ACCUMULATED KNOWLEDGE OF THE AGES

A Special Supplement of the most fundamental knowledge gained by me in my eight years of therapy, 7 years within the A. Justin Sterling organization, The Nation of Men, The Forum, Lifespring, the Ropes Course, two Alumni Reunions, countless team meetings, several Community Service Projects, multiple relationships, Open Houses (big and small), Anthony Robbins Unlimited Power and ... (With credit and/or apologies to those I quote here).

- No matter what a woman asks you, what she wants to know is "Do you still love me?"

- Scream and Leap.

- Trust the Men.

- Some Will, Some Won't, So What ... Next!

- If you find yourself doing a transfinite (a Lot) amount of work the answer can be obtained by inspection.

- Instead of looking at a task and asking yourself "Can I do this?", ask yourself is there anyone else to whom I could delegate this task. Look for ways to not DO. That's leverage and it's one of the major differences between those that succeed and those that don't.

- If you have a fifty-fifty chance of being right you will be wrong 90% of the time.

- Given one obviously wrong approach to a problem and one obviously right approach it is often better to choose the wrong approach thus forcing the inevitable breakdown, sooner.

- Success leaves Clues. If you want to succeed at something find someone who has succeeded and do exactly what they do.

- Whether you think you can or you can't, you're right.

- Wherever you go there you are.

- Life is a mountain without a top. The trick is to learn to love the climb.

- Every action you take has a destination associated with it; the trick is to know *your* destination and take only those actions which lead to it.

- The quality of our lives is governed by the quality of questions we ask ourselves. The trick is to learn to ask yourself better questions. If you ask yourself "Why am I so stupid?" your mind will find a very good answer but this adds little to your life. To get more from life start asking better questions.

- The hurrier I go, the behinder I get.

- The earth sucks. If you wait until all of life's details are handled before you go for the brass ring, you never will. Go for your passion and life's details will take care of themselves.

- Life is simply the pursuit of Faster Cars, Older Wine and Younger Women.

- I've been rich and I've been poor. Trust me, rich is better.

- Money may not solve life's problems but when you are rich at least you can arrive at them in style.

- Speak only with good purpose. Anything else is just noise.

- You can get more done with a kind word and a gun than you can with a kind word alone.

- A man thinks his woman will never change, a woman thinks her man will. Both are wrong.

- Life is a learning process. You learn more from failure than from success, so fail often and well.

- There is only one important thing in life. You just have to figure out what it is.

- Birds can fly but fly's can't bird.

- Water is wet and the sky is blue.

- Sears has the best tools.

Baseball at Thirty-three and a Third

BILL FRIEND
AaaRrrrr! ♂ *SBNoM*

Nearly every day, my four-year old son Alex asks me to play baseball with him. I am amazed at his interest and enthusiasm at such an early age, but then maybe I shouldn't be; after all, he viewed his first baseball game on TV when he was only 10 hours old (while wearing an Oakland A's helmet — actually an ice cream helmet cup from the Oakland Coliseum). At three weeks old he went to his first A's game. By the time he was five months old he had attended 10 games, including Game One of the 1989 World Series against the San Francisco Giants. I figure having all this baseball crammed down his throat at an early age will either turn him off to the sport (it still may), or he will develop a passion and interest that we can share all our lives, and he can pass that on to his children and their children.

Baseball has been a part of my life almost since I was born. It's been my passion for over 20 years and continues to be even now as the game changes, and the ball players have become travel-ing mercenaries — willing to play for the highest bidder, in any city, for any team. As I grew, baseball wove its way into the shaping of my being and philosophies on life.

In 1960, the Washington Senators announced that they were moving their franchise to Minneapolis/St. Paul — in fact they would play only a few miles from where we lived in St. Paul. I was only a baby. Although I can't recall the excitement and enthusiasm of having Major League Baseball move into our city, I am sure I was aware of something special going on. It is interesting and poignant to see our family's first color home movies showing my oldest brother Bob pitching a baseball to me. I'm 2 years old, wearing a Minnesota Twins cap,

suspenders (no shirt) and baggy pants (wearing no shoes—or for that matter, no coordination).

The Twins captured the hearts of the upper Midwest in the Sixties, and my family was no different. Bob and my sister Judy went to dozens of games and witnessed some unusual moments such as two grand slams in one inning and five home runs in one inning of another game (against the *Kansas City* A's no less).

My dad watched all the televised games and listened to most of the others on the radio. Although I don't remember any particular game or play, I do recall the closeness and, from the perspective of today, the bonding I shared with my Dad whenever a game was on. I vividly recall the oppressively hot, muggy nights we watched games on our black-and-white TV — its glowing screen the only light on in the room as I laid in front of the noisy air conditioner that tried its damnedest to cool us off. I remember my other brother Jim and I laughing hysterically whenever my Dad would let off steam by swearing at the umpires, the other team's players and their good fortune, and even any of the Twins players who *had the audacity* to strike out, make an error or give up a home run. I'm sure it was my first exposure to such colorful words as "son-of-a-bitch" and "bastard."

I wondered aloud as to how they could play baseball at night since I couldn't grasp the concept that there were lights bright enough to light an entire baseball stadium. Jim (who I think stayed up at night thinking of ways to mess with my head) explained that the games were actually played in China where it was daytime and beamed back to the states by satellites. *That* I believed — it made sense. Of course I also believed him when he told me that Brooks Robinson and Frank Robinson of the Baltimore Orioles were brothers (only one of them was a black man ... in fact, I believe he still is).

The Twins made it to the 1965 World Series, but lost to the Los Angeles Dodgers in seven games. Although my Mom and Dad rarely went to games in person, I think they were disappointed that they weren't able to go to the Series, either because they couldn't afford the "over-priced" tickets or because most of the available seating was typically bought up by season ticket holders and large corporations. That is something I've always remembered about that time.

By 1968, baseball was not a big deal in our house anymore. Bob was in Viet Nam, Judy was graduating from high school and we were all packing up to move to California. Baseball made some rule changes that year that made it more of a pitching game than a hitting game with low batting averages (Carl Yastremski was the only .300 in 1968). The excitement for us had worn off and it seemed baseball was now boring and trivial.

Also packing their bags and moving to California in 1968 were the Kansas City Athletics, a fact that we noticed or cared little about until the 1972 World Series. By then the *Oakland* A's were a young, brash, colorful, long-haired group of players facing the confident, conservative Cincinnati Reds. It was billed as the "Hairs vs. the Squares." For the first time in years my family watched baseball together. It was all new to me, the rules, the nuances of the game — even who the hell actually played in the Series (American vs. National Leagues). My Dad explained them and reminisced about the Twins and other games he had seen and heard. My interest and excitement grew with each game until that last out was made, and the A's were World Champions and, in the

process, had re-ignited the passion of baseball and the passage from father to son. I was hooked.

It was the best time to get hooked by baseball if you were an A's fan because they won the next two Series and it seemed they were invincible. Jim started taking me to games, and we created memories and experiences together just as Bob and Judy had done ten years before. It was a special time in my life as I could escape the insecurities and self-consciousness of my early teenage years by tuning in the ballgame on my radio and listening for hours. I dreamed of someday traveling to the old historic ballparks that I heard so much about: Yankee Stadium, Comisky Park, Wrigley Field, Tiger Stadium and my ultimate fantasy — Fenway Park in Boston.

Baseball was the greatest thing in the world! "My" team had won five straight division titles and three World Championships. They were young and nearly invincible. And then I found out the truth about baseball: it was just a business. On June 15, 1976, with the trading deadline only hours away, Charlie Finley, the A's flamboyant owner, **sold** his best players: Vida Blue to the Yankees for $1,500,000; Rollie Fingers and Joe Rudi to the Red Sox for $1,000,000 each! It was the year of the first wave of player free agencies, and Finley believed (with good reason) that all his players, players he spent millions of dollars developing in the minors, would simply leave at the end of the season, leaving him without anything in return. He had already lost Catfish Hunter through free agency and had traded Reggie Jackson earlier in the spring anticipating his free agency departure. I had never felt so devastated. It was a wound that slowly developed into a callus, a callus that would try to protect me from future disappointments. Years later, I would feel that pain as players such as Rickey Henderson, Tony Armas, Jose Canseco and Dave Stewart were either traded or left as free agents. Whenever I'd start feeling that breach of loyalty I'd tell myself "it's just a business," but that old pain from summers ago has never really been completely suppressed.

Although Finley's sale of his ballplayers were nullified by the Commissioner of Baseball, the nucleus of his championship team left for other teams just as he had predicted. The A's fell into the cellar just as I was old enough to drive myself to the Coliseum and, despite their ineptitude, I went to dozens of games and sat wherever I wanted to amongst the thousands of empty seats. Many times I would go to games alone, and in doing so discovered that baseball still let me escape the worries, insecurities, frustrations and even the loneliness of my life. By then it seemed I was always pining for the affections of one girl or another who would "*treasure our friendship too much*" to ruin it with a relationship. Baseball kept me company, and it didn't mind if I was obsessed or impassioned about it.

In 1980, I actually had a real live girl friend and I took her to a number of games early that season. May 15, 1980, was a day that I always remember whenever I think of Renee. It was a beautiful, warm Thursday afternoon. Although we had been dating for a few months, I had never really let myself "fall in love" with Renee — after all, I was still pining for the unattainable girls that I couldn't have but wanted. That day she was more beautiful and more affectionate and attentive than I had ever seen her! No, we didn't have sex under the stands or anything like that (in fact we never had sex in all the time we dated ... Geez, what the hell was I thinking?). Something happened that day that made me finally fall for her, and I don't quite know what it was. Perhaps it was the letting down of my shield that hid my insecurities and the sharing of that special place with someone who actually wanted to be there with *me*. It was such a perfect day. It was how I always thought having a girl friend would be like, filled with humor, affection, pride and self-confidence — and yes, baseball. It was a long game that the A's won on

Tony Armas' grand slam in the bottom of the 15th inning. And we stayed there until the last pitch.

By the end of that summer, Renee started dating someone else and that pain of losing my first girlfriend was *almost* as painful as that summer day in 1976 when Finley had sold his players. I felt betrayed and lost, but baseball had prepared me for my heartache and was there to soothe it.

I drove to Seattle in 1982 with my friends, Tim and Larry, to see the A's play in the Kingdome. The A's also played an exhibition game in Tacoma against their minor league club, and by being in the right place at the right time, I found myself *inside* the A's dugout for the entire game. Before the game, while shooting photos from the stands down the left field line, some executive from the Tacoma club asked me to shoot a photo of their mascot for the local newspaper. I walked onto the field and shot a photo of this stupid looking tiger — then he suddenly took his head off, and there was A's manager Billy Martin inside! I was stunned, and unfortunately was unable to get a good shot of him, as he appeared and left so quickly. The game started almost immediately. I was standing in front of the A's dugout and since nobody asked me to leave, I sat on the end of the bench, excited and thrilled. I was like a fly on the wall or, perhaps more appropriately, tobacco juice on the wall as I watched and listened to these guys. I shot all the rest of my film, and by the fourth inning, I decided I needed to get more film from my friends up in the stands (who at that point didn't know where I was). I left through a gate and told the security guard that I would be right back. Tim and Larry were skeptical when I told them, but watched as I went down again between innings and security let me through the gate and into the dugout.

Those photos led to an opportunity of being a contributor to the A's magazine the next two years. As exciting as it was to be down on the field at a number of A's games, I soon found out I was losing my reverence and innocence for the game and its players. I came to realize that these players were just like you and me — well, that is, if we were millionaires with egos the size of our well padded wallets. They weren't gods or heroes, they were just guys, many of whom seemed to be arrogant jerks. I felt like I had just discovered the truth of the Wizard of Oz ... *"Ignore that man behind the curtain!"*

By 1985, I found myself back in the same old rut of pining for the affections of someone I had put up on an unattainable pedestal. I was miserable. I had just graduated from San Jose State and was still living at home with my parents. I was eager to travel, grow, learn and experience some independence. Most of all, I wanted to get away from the reminders of all the heartache, loneliness and self-pity I had let myself become victim to. That spring I planned the fantasy trip I had dreamed about since I was 13. I was going to travel alone across the country and visit various cities and stadiums. I jokingly refer to this period as "the time before I got a life"; in actuality it was the time that I started to "get a life," a life of confidence and maturity as it set me on the path to self-discovery that led to my growth as a man, and ultimately husband and father. I recommend to everyone that they plan and take part in a big traveling adventure, preferably alone, revolving around whatever your passion is. It's amazing how much someone can grow from that experience.

It took many hours to schedule the best time to go on this excursion since I would want to make sure each team was in town on the date I planned to be there. That August I flew into Minneapolis and was greeted by my cousin Jean and her husband Jeff, also baseball fanatics. They took me to the new Metrodome and then to Chicago where I would start the rest of my journey alone. We took the elevated train to Wrigley Field for an afternoon game and drove down to the south side for an evening game at old Comisky Park. These stadi-

ums were like cathedrals to me. It was nearly a religious experience as I walked through these old, quaint, charming ballparks. I just couldn't get enough of them and returned for another "double-header" the next day after Jean and Jeff had left for home.

The unbearable mugginess in the Midwest immediately took me back to those hot summer nights as a child in St. Paul. This time I didn't mind the humidity or the mosquitoes, in fact I considered them part of the ambiance: baseball, hot dogs, and sweaty bites.

I left Chicago the next morning, and it finally hit me that I was on my own. As I drove east toward Detroit and into some angry looking storm clouds, I could feel a heaviness behind me as my whole life — my family, my friends, my job — everything I knew was 2,400 miles away. I was heading in the opposite direction and it seemed so unnatural and a bit frightening to be so far away from my world. I had visualized this trip for so long that I felt as though I was driving on this gigantic road map that you'd get from a gas station, except instead of a scale of 1 inch equaling 10 miles, this one was more like 1 mile equals 1 mile. It was almost as though I could see that Texas was to my right, New York and Boston were straight ahead, and in my rear view mirror, just over those Rocky Mountains, was my lifeline.

Detroit turned out to be quite a test of for me. Arriving late I decided to find Tiger Stadium and rent a room nearby — not the brightest move since the area I ultimately wound up in was pretty rundown, and the only place I could find open at that hour had steel bars in the windows and doors and a parking lot reeking of urine. I was tired and sweaty and I didn't care, I just wanted to take a shower and go to sleep. It was nearly 1 a.m. by the time I sneaked in all my camera equipment and luggage. I soon found out that the room had no towels or soap and the bed wasn't exactly the freshest smelling. It occurred to me that this

METAMORPHOSIS

motel's clientele probably paid by the hour. What the hell, it was a place to sleep ... or so I thought. At about 3 a.m. I was awakened by a huge crack of thunder as a midwestern storm passed over me. As I stood by the window and watched lightning touch down less than a 100 feet away (*kids, don't do this at home*) I got a good view of what a sleazy and dangerous part of Detroit I was in. I wondered if mothers of small children in that neighborhood soothed their frightened children back to sleep by telling them that the thunder was only the sound of gunfire.

The next morning I checked into the Hyatt Regency in nearby Dearborn as a reward for surviving the previous night's ordeal and in anticipation of a long day waiting to see if the constant drizzle would rain out that evening's game. The game *was* played, and Tiger Stadium was worth the wait. A beautiful old ballpark, Tiger Stadium was a sea of vibrant blue, orange and green. The whole structure was painted deep blue with vivid orange seating and bright green grass glowing under the misty lights. It is definitely a park I want to visit again.

Two days later, I spent 7 hours wandering through the Baseball Hall of Fame in Cooperstown, New York. I had been there before with my family when I was 16 but being there alone enabled me to spend as long as I wanted to at each and every exhibit. I will someday return there with my own family.

In New York City, I finally got to see Yankee Stadium, a ballpark so steeped in baseball history that they have their own little Hall of Fame beyond the center field wall. Before they refurbished this old ballpark, the monuments actually were *in* center field, and any ball hit out there was in play. I loved this park and wished I could have seen it as it was before it was modernized in the early 70's.

As much as I liked Yankee Stadium, I absolutely hated Shea Stadium where the New York Mets

play. I found it ugly and rude. Perhaps it was the long drive out to Flushing Meadows or the terrible seats I had; whatever it was I couldn't wait to get out of there.

While in New York, I took in the typical tourist sights — Statue of Liberty, Empire State Building, Times Square — all in humid 95° heat. It was in Times Square that I came to realize that there is something about the smell of HOT urine that is really repugnant. Oh, and the sights — all I could see was sex, sleaze, drugs and filth and that was just at the McDonald's I stepped into! Jay Leno once said, "If God doesn't nuke Times Square, I think he owes Sodom and Gomorrah an apology." I think he may have been right. Yet I wasn't intimidated or frightened for my life; in fact, I loved the excitement of this city that never sleeps (or bathes). By now, I had been away from home for over two weeks, and I found myself feeling very confident and in control of my life and destiny. My self-esteem was growing each day. I found I was not looking forward to heading home just yet.

After a sight-seeing trip through Philadelphia, Baltimore and Washington D.C., I headed back through New York a few days later and on to my ultimate goal — Boston's Fenway Park. Wouldn't you know it, I traveled 3,000 miles and half my life to get to Fenway, and the game got rained out! After 2 innings of play in which the Minnesota Twins hit back-to-back triples off the "Big Green Monster" (the 30' wall that separates left field from the street behind it) the game was postponed for 3 hours. I didn't mind though because I spent the time wandering around Fenway memorizing every angle, every sound, every smell of that beautiful, intimate ballpark. For years I had dreamed of walking into Fenway and standing in those last left field seats next to the "Big Green Monster" and touching "The Wall." I am not a very spiritual or religious person, but for me it was like visiting Mecca. I stood in that spot I had dreamed about, and it was exactly as I imagined it!

METAMORPHOSIS

Only three and a half weeks had passed since I had left home, and yet I felt different inside. I let go of emotional baggage that had been mind-talking me to misery since junior high. I gave me a better image of myself and found I really enjoyed spending time with ME. I admit it did feel good to have friends happy to see me again —I felt as though I belonged and was not some outsider in my own world.

Baseball called me back East again the next spring as I did a mini-baseball pilgrimage over a numbing four day—four city—four game period. I flew into Minneapolis and immediately drove with Jean and Jeff to Chicago to see the Cubs in Wrigley Field on a Friday afternoon. We drove to St. Louis the next morning and caught the night game at Busch Stadium. Early Sunday morning we drove to Kansas City for an afternoon game in Royals Stadium, where we saw George Brett's 2,000th hit. The temperature was very hot and the game went into extra innings, so we had to leave in the ninth inning for our return to the Twin Cities. We were already in Iowa by the time the Royals finally won in the 16th inning. By Monday afternoon, I was burnt, yet I wanted to see one more game at the Metrodome before I flew out that night. I'm not sure what significance this trip had for me other than I was going a bit overboard on this baseball thing and I really needed to "get a life."

That year, I took the Sterling Men, Sex & Power Weekend, as well as the Life Training Weekend. They both embellished and strengthened my newfound confidence and self-esteem and cleared out more of the garbage I was carrying around in my head. I believe it's no coincidence that as soon as I learned to love and respect myself I met someone who ultimately was THE ONE. Less than two years later, I met and married Anne, who was nothing like who I thought I was looking for — yet was everything I ever wanted in a wife and mate.

By 1988 I finally became a season ticket holder of the Athletics with seats behind home plate. The timing was good to reserve these seats as the A's became a dominant force again in baseball over the next several years. My fondest memory of the A's return to the top was the day they won the American League pennant for the first time since I was 14, 14 years earlier. I had tickets to that pennant clinching game and, believe it or not, I was unable to attend the game or even listen to it on the radio because of a prior commitment I made with Anne (the first *real* test of our new love). I had a whole list of friends who would gladly have paid twice the face value of each $40 playoff ticket I held. However, the decision I made I made without any hesitation — I gave them to my Dad and Mom as that day was also their 43rd wedding anniversary. Knowing they were able to witness a championship game in person and share in the excitement and electricity of being there was something that I was glad to provide and be a part of.

It's ironic that I married a woman who doesn't share my passion for baseball, yet I don't mind. She makes her one appearance on opening night every year and gets her fill for the season. Over the last few years I have found myself going to more and more games with my sister Judy. We spend the whole time talking about baseball — she of her fond memories of the Twins in the 60's and me with my memories of the A's in the 70's as well as our shared memories of the current A's of the 80's and 90's. She has passed that interest to her son as I have to mine. After all, baseball isn't just a game for fathers and sons — it's also game for fathers and daughters, mothers and sons.

By coincidence, my daughter Alyssa was 6 months old on Opening Day this year. You can probably guess where I took her to celebrate.

EXPRESSIONS

METAMORPHOSIS

METAMORPHOSIS

Longing

CHRISTOPHER DINGMAN
Sperm Whales ♂ SCMC

I long for you like a river that stretches out ...
Like a man that walks down a city street.
You are like my childhood.
I think too much without you.
I will go home now
and bandage my heart
with dreams.

ORIGINAL WATERCOLOR BY
PETER ASHLOCK
Wailers ♂ EBNoM

To the Woman the Man Loves
A Gift for Deborah

GLEN BARLAS
Spartans ♂ EBNoM

To me you're a rose
Your beauty is subtle
It emanates from deep inside your heart.

When I touch you or am
Touched by you, I am refreshed
By your feminine soul.

As my heart meets yours
I am held by the love
that binds them both.

Beloved One, From the Forest

DAVID RUBINE
Rod Warriors ♂ SCMC

Beloved one, from the forest.
What image do they continue to duplicate?
An unshaven face, no beard masks the tide of feeling that moves,
old — a man sorting through the triumph of his memories.
What, then alone, the green shroud escapes notice,
excepting the sun which burns age into the supple bark.
Moss, damp darkness, the alien certainty of fungus, what crawls
beneath the ragged feat.
The earthly kiss of night, more silent than a whisper, beckons the old one home.
The sentient veil presses its soul fingers into the loam,
deep prints awaken the memories of a forest who remember his passing.

Transformation ... Ongoing

BILL McCONNELL
Phoenix ♂ *SBNoM*

Transformation is very elusive. It comes and it will go unless we maintain our own discipline to keep it happening. It takes a lot of hard work, and it never ends. The path that one takes to realize transformation within his life is really the experience. Transformation, as such, is kind of "way out" there for most people. The interesting thing about taking this path is that one dare not attach to the idea of transformation. If a person does attach to it, transformation disappears.

Man's religions have at their core the idea of transformation. Any of these religions, if one is really interested in going down the path, say transformation is a way to have it. The path is filled with potholes, twists and turns, and even demons. These demons appear all the time on the path; one can easily become distracted by them. That is why one needs to be conscious and able to concentrate on his experience of the path rather than the goal.

One has to be willing to change his whole notion about himself and life. I think greater courage is needed on this path than to face death in a battle. There are demons; there are temptations that appear; one can easily be seduced. The path is full of ruts, heavy rains, fog that seems impenetrable. This and the demons really test one's commitment to one's self. After all, this is what transformation is: realization of who one is.

I have a dear friend whom I am going to talk about. He held a powerful executive position in his company. He was married for the second time; his wife and he have two children. This man has a law degree and a degree in mechanical engineering. Suddenly he was let go. All the middle management of the company were let go in one swish of the pen by the president of the company.

When the word came out that he was out of work he was shocked; he did not think that it could happen to him. He became depressed; he was asking serious questions of himself. When he looked he began to realize that he was doing things in his life to please others and not himself. He also realized that he had lived in a dysfunctional family, and for him to survive and look good in that family he had to achieve; he had to become the best in everything that he undertook.

As with any loss, he had to learn to grieve that loss. Who was he now that he did not have this power job? Who was he to his family, to his friends, to the men in his life? How could this possibly have happened to him?

He called me. One of the things that became clear to me by the questions he asked was that he was looking at what to do now, where to go. He did not like being home because there was so much silence, so he needed a person like myself who would listen, upon whom he could bounce off ideas.

Naturally the loss of the job created a lot of strain on the relationship. His wife had to find a way to support him and at the same time put up with his changes, because he was going through many changes, huge emotional changes. She showed a great deal of patience, and even thrilled at what her husband was going through. Suddenly he seemed like the man she knew he was all the time, but this change meant that she was going to have to go through changes herself. She was excited, yet frightened about what all this meant.

We spent a lot of time together during the first part of his experience. I suggested that he get to a hospital and get a physical and also think about seeing a therapist because I was not a professional. He found that physically he was in pretty good shape; however the anxiety that he was going through was difficult and causing him to have physical things happen that bothered his sense of well-being.

He went into therapy where he learned that he was very depressed, that he needed to work through the depression. Soon he was looking at his life and beginning to work things out as to what he wanted in his life and how he wanted it. This was the first time in his life that he realized that he was important to himself. I suggested that he hand a lot of this stuff over to his Higher Power, and allow the Higher Power to support him in this and in taking all this on. He said that he was never a religious person, but he would try this. He did this every day for a while and awesome things began to occur. It was a thrilling experience to see what was going on within him. He was finding out that he had to honor himself, he had to honor the people in his life, he had to give of himself in a way that people in his life would experience this.

Although he put out many resumes and filled out many job applications, he realized that he was really not interested in doing that, but rather in giving to his family, to his mother, his sister and brother. He realized that he was the person who had to create this, he also realized that he could not do it by himself; he could do it, however, with his Higher Power's support. This change in him meant that his wife had to look at what she wanted

in life. He wanted his wife to have her dream, too, because having his dream without her having hers would be awful. She had to choose to be with him, to re-choose, more or less. He asked her what she wanted, what was her dream for herself and her life, could they do it together?

He had become very vulnerable emotionally. He was questioning his motivations: why had they shown up as the did in his life. He realized that he had not really lived an honest life, and he had not been sensitive to his wife's needs, and he realized that he had not allowed himself to be vulnerable to his wife. You could tell that she really liked what was going on; however, she was confused. Suddenly she was feeling appreciated for all that she is. What was her dream? It was elusive. If she went for what she thought she wanted to do, it would require a lot of commitment and expense. Did she have the commitment, did she have the strength, the willingness? For her it has meant that she had come fact-to-face with her own experience of herself. Could she really have this become her dream for herself? Did she feel well enough about herself to take advantage of this opportunity? These are difficult questions that we all have in life, questions that many of us are afraid to look at, and when we do not look, we miss out on having our own experience of transformation.

They have chosen to do this together. Many a wife would have chosen to leave, many a man would not have felt well enough about himself to take the risk that he is taking. He decided he wanted to be able to give back to the community, so he is getting his teacher's credential. He is also building an add-on to his mother's home with the intent of living there with his family, so that they have separate living quarters, his kids could have an experience of their grandmother, and she could experience them. He is so very involved with his process, his wife with hers, that when I have seen them recently they looked the best that I have ever seen them. It is just wonderful to see how this whole process is working. Both seem to have a very good experience of one another that shows their willingness to support one another through a tremendous change in their lives.

Each day, each week, have been transformational experiences because they have been willing to support one another in having their respective dreams for one another. He realizes now that although he held a very prestigious job with a company, he did not have in his life what he has now. He realizes that he was living a kind of an insane life then, but now he is very alive, he has made his choices, has been able to support his wife's choices and is continuing on his quest. For them transformation is occurring every day in their lives and it is not easy, but doing what they are doing has become a way of approaching life, and if they can keep that self-discipline going they can have all that they need and want for themselves out of their lives.

It is clear to me that transformation has to do with being one's best. It means learning to be as creative, as positive as possible in all one's encounters, even though there are things going on in the world and in life that are destructive. One must have the courage to withstand the slings and arrows of life and do it in in a way that reflects one's joy in all that is in the moment, in the only **now** one has. Even though all kinds of things are happening, the discipline is to learn that one is okay just as one is. This is not easy, no way is it easy. We do have to be within our moment of the **now**, sometimes be with very painful feelings, but not act them out against anyone in our own universe (our family, our spouses, our children, our acquaintances, anyone that we know of or can know). We have the choice to keep our awareness such that we can always be on the edge of creation, of contributing to people in our lives and to ourselves.

Realize that this moment is all we really have. We cannot seek transformation as a goal in itself. It is the path leading to it that is important. This path,

full of all kinds of twists and turns, even demons, is the path we must take. We will be confronted with our negativity and we have to make the choice as to what this means in our lives — do we want to dwell upon it; do we want to release that into our lives and our universe? We do have the choice in the matter. If we choose to be positive, it means to be active, to be willing to take chances and face the changes that come along without allowing those changes do anything to our faith in who we are. This is what it means to be spiritual: to seek direction from a Higher Power, to be grateful for what that Higher Power has given us in our lives.

All of this is tranformational and is a part of the long path that one must walk. We can choose to have a change in our attitude towards anything in life, in one moment, in this very moment. Changing the attitude is one thing, keeping it so that it is **the** attitude that we maintain is another. We can slip back easily and quickly.

We, along with our Higher Power, need to be partners in this process so that we not only honor ourselves, we honor those in our lives. We have the choice. We find that there is much that will come up that is so very painful. Going through that pain will give realizations that are awesome and will allow one to experience oneself in a wonderful way. If we learn to love ourselves and our Higher Power, all will be given to us. The question is, are we willing to head down that path? It is full of potholes; there are demons that will attack us; there are twists and turns and even Y's in the road where we have to make the decision to go one way or the other.

Are you ready, are you willing to persevere, do you have the faith, do you have the commitment? Are you ready? Then, let's get on with it ....

Self-Portraits

JEFF GODFREY
Spartans ♂ EBNoM

A while back I had been keeping a journal to jot down feelings and experiences I'd been going through at the time: bicycle trips, thoughts about my job, and a lot of who I was in love with this week and wouldn't-it-be-great-if-we-were-together stuff.

These eight drawings took the place of journal entries during a period when writing down words seemed futile. This was when a woman who had been a friend and a workmate for some time let it be known to me that she could no longer suppress her feelings. In the form of poetry and artwork Jane handed me a booklet filled with passion and adoration. She expressed parts of her I had not seen before. She was in love.

I was blown away. Last week we were good friends; now I was the sun and the moon and the stars to her. I thought she wasn't even interested in men. So many rushes of mixed feelings running through me. I didn't know how to respond, so I didn't. I drew. I contemplated what this all meant to me. I burnt inside, because even though I'd been looking for a relationship with Diane or Cindy or Yolanda or Grace, Jane was not someone I had thought was even an option. I did not choose her.

A week later we did talk and decided to go on some dates. On our second date we went to dinner and then hottubbed together. It was not what I expected but it was exciting.

Over the next week everything began to change. The barriers I once had to her dissolved. The longing feeling I carried in my heart was replaced with contentment. Again, drawing was the best way to express my feelings.

While I feel I did not "choose" this relationship, I don't think I could have chosen a more perfect one. Years later, Jane and I now have a beautiful relationship, marriage and daughter. I feel so fortunate for what I have.

ORIGINAL PEN & INK DRAWINGS BY
THE AUTHOR

Prometheus' Fall

RON DEMATTIO
AaaRrrrr! ♂ *SBNoM*

I am a thief in the night.
The night is the darkness in me
from my faulty training.
I protect my wounds,
they become the focus of my soul.
My soul cries out, "release me, it is
only a wound...
nothing to die over!"
I envy you your completeness
you have no wounds that **I** *see*
I envy you that which
I think *I see.*
With my envy, I steal from you
your humanity.
You are not allowed
to be less than enviable
and the fire is stolen
from both of us.

Unity

WAYNE GILBERT
Guest ♂ *SCMC*

When you think about it,
Sacrifice and pain ... is an everyday thing
For people gettin' along, a little at a time
When you think about it ...
We all need a little help ... we all need a little change
But too many good people are brought down
by the prejudice and hatred ... all around
Now's the time ... for the righteous to unite
And not slow down ... till the whole world's put right
When you think about it ...
Why can't we all work together
And maybe for a moment
Stop all the fighting ... and stop all the hating
Let's put inside ... a child's tear
a little love ... where there used to be fear
Let's start the future right now
by me helping you ... and you helping me
When you think about it ...
We can't live like this ... anymore
Let's take that step ... through the peaceful door
For all the lonely ... and all the old
For all the innocent ... left out in the cold
Let the righteous unite us ... and help all of those
When you think about it ...
We should do something about it.
We should do something about it.

Fossil-Hunting

John Renish
AaaRrrrr! ♂ SBNoM

We cross the desert plain,
nomads seeking neither water nor pasture,
but stone,
where seas once swelled warmly,
turgid, amniotic, cradling life.
We stoop to gather, gleaners in a field of sand,
a scallop, a snail, a scuttling trilobite,
a broken something-I've-never-seen.
A fragment of ammonite squirts from under my boot
as a watermelon seed slipslides through a child's fingers.

The Collision of Power and Beauty

David Rubine
Rod Warriors ♂ SCMC

The collision of power and beauty wrought an effortless, effulgent
light the day you were born.
Constant and burning, a fiery beacon that shines for all eyes,
your radiance has come to know no discretion.
In that light we walk with our own shadows and know something immense
has lit our way.
You are that immense.
What measures you measures the world!

We try to know the bounds of light, the circumference of love and watch
you love him like the world was made for God,
his kingdom is your body, his justice is your soul.
Loving the world is like loving him whose soul broods about eternity
and laughs at the seriousness of men.
Do you know who you are, who would be willing to love the world forever?

When you leave the world of men and enter that nameless sanctuary
don't stop defying the voices that have tried to name you.
Your name is what trees whisper and streams lightly sing.

METAMORPHOSIS

Transforming Guitar Playing

Spartans ♂ EBNoM

"Joplin in Rio" was written quickly, composed on the guitar in a couple of days and then revised several months later. The music captured a state of mind resulting from receiving some personal and painful news, and it marks a turning point toward freedom.

About 16 years before I wrote this piece I had a transformative experience, involving the guitar and a large dose of LSD, that has colored my life since. This was a multicolored and multidimensional "gates of heaven" sort of experience, and I was pretty pissy for a few years about having to come back and live in the real world.

When the karmic dust had settled, I still had one thing that I had learned and brought back, a very simple trick about playing the guitar. This may be expressed as: Each finger of the left hand must come down firmly and precisely on the point just in back of the fret and must under no circumstances touch either of the adjacent strings (whether they are vibrating or not)!

Try it, you'll like it!

Leaving out discussions of whether this is so obvious it's dumb, or too hard to be worth it, or whatever, let me say that (being a klutz myself) this little principle has given me something to work on for the last 18 years.

ORIGINAL MUSIC COMPOSITION BY THE AUTHOR

151

METAMORPHOSIS

Wesley

TIMOTHY BUCHANAN
Golden Gate Jerks ♂ SFT

This picture is a portrait of Wesley, a good friend of mine. It so richly describes to me the feeling of misery and emotional pain.

ORIGINAL PHOTOGRAPH BY
THE AUTHOR

METAMORPHOSIS

What It Is

SHAWN RILEY

More Power! ♂ SBNoM

It is impossible to communicate the male experience in words alone.
The men's movement is.

Change

HOW SHAW

NFL ♂ SBNoM

Courage

Head-on

Acceptance

Now

Growth

Energy

A Review of the Life Training

SAM ZUCKSWERT

AaaRrrrr! ♂ *SBNoM*

I attended a profound, life altering experience known as the "Life Training" that may have been the most valuable experience of my life. This weekend event was deeply experiential and awakening. It Blew my mind, Opened my heart and Strengthened my will. I benefit every day of my life from this transformation.

Do you recall what it's like to really know something from direct experience rather than understanding it intellectually? Those things you know from direct experience have a "Knowingness" and certainty that resonates from every cell in your body. Intellectual understanding compared to knowing by direct experience is a real booby prize.

An example of this difference would be understanding the experience of an **orgasm**. You could understand the physiological processes in the body and measure the frequency and size of body contractions and on and on; but this "understanding" is so preposterously alien to the actual experience of an orgasm that it's laughable!

Any description of the Life Training experience suffers this same impossible limitation. If you've taken the Sterling Weekend and tried to describe your experience and benefit to someone else, you probably know what I mean.

The Life Training experience is so profound that it is especially difficult to explain. Its stated purpose is:

The purpose of the Life Training is to awaken in you an experience of life that will radically enhance the quality of everything else in your life.

I will attempt to give you a sense of this awakening. You will learn from direct experience things that you absolutely didn't know about yourself. You'll learn how you have been unknowingly making your life miserable. After knowing how you did it, you can choose to get off it!

The Life Training will give you a direct experience of life that is so different, enhanced and profound that you will never again be the same. After knowing that this different domain of living is possible, you will be able to create a new life filled with aliveness and possibilities.

One of the principal differences between the Life Training and other "growth" experiences is that you learn specific tools that you can use every day to enhance your life. With these tools I have learned how to discover and eliminate the unconscious lies that my mind uses to limit me, put me down and make me dissatisfied and unhappy.

I learned by direct experience how to open my heart, without throwing my shield and self-determination away. Like almost everyone else I know, I had many painful experiences in the past that led me to tightly close my heart. I closed my heart because I falsely believed that this was the only way I could be safe and survive. My closed heart cost me the most deep and meaningful part of my life. It cost me love, joy, sex, and passion for life. Through Life Training I learned how to let my heart open and expand, to feel joy and passion for life. My heart sings a lot more now and I'm truly grateful for Life.

Life Training tools showed me how to give up the tremendous pain and cost of resentment without having to go back for a second dose or condone what happened. When I am feeling resentful about something or someone, my jaw is clenched, my vision is tight and narrow. My heart is so bitter, tight and closed that it only feels numb. When I am in this disease of resentment, it doesn't matter how beautiful the day is or what great things are going on, my whole life is just being the experience of resentment. I am now able to quickly stop this self-made hell while maintaining my shield and free will.

I learned how to strengthen my backbone with the inner strength of Clint Eastwood's "Man with No Name." My strength now comes from self-esteem and true confidence, not from anger and defensiveness. I defend my territory with confident, strong, self-assured ownership, not with permission, justification or apology.

Of all of the trainings I've taken, including Sterling, the Forum and est to name a few, I value the Life Training head and shoulders above all the rest! Many SBNoM members have taken the Life Training. It really works — ask them about it. You can get further information about the Life Training by calling (510) 651-0270.

Becoming New Again

DON ADAMS
Guest ♂ SBNoM

Searching inside ...
for the best left aside
Clearing today ...
yesterday's barricades away
Setting new limits ...
without being timid
Stoking the fire ...
that sparks and inspires
Revisiting a forgotten dream ...
refining it with the team
Becoming real brothers ...
losing my druthers
Traveling honor's road ...
sharing with the load
Honoring each man ...
becoming new again
With honor and integrity ...
we celebrate our victory!

Kala Lau

STEVE BACKINOFF
Chi-Pachagon Express ♂ SCMC

This poem is in honor of my parent's 45th anniversary. The artwork is my son's, Tim Norton.

I remember the double rainbow

                    over the deep green valley
      over the orchard where we made love,
                                filled with hope that all people could know this peace.
But
      I still see red,
                        angry growl in my gut
                                          covering my grief
                                                    Well of tears
            knowing that this wonder-filled life is full of loss.
      Why me?
            Why not me?   Why did God take her?
            How could they not know
                              how badly we all wanted to help?
      And not only cancers and suicides
                                          but the little deaths
            teeth,   hair,   knowing that it is less & less likely
      that I will ever father a child
                                    and watching loved ones not make
            the healthful choices   that I see so clearly.

      With all of this weighing
                              We still get dressed up
                                          meet with friends
                                                    and go to the dance.
            the barn,   hands round wrists
                              twirl   the human star
                                          your partner   do-si-do.
      Memories of pagan leaping fire
                              naked poems of life   licking the sky
      illuminating the forest standing round.

METAMORPHOSIS

Original watercolors by
Tim Norton
Guest ♂ EBNoM

We find our way home   with beloved   to share the hearth
          reaching out   from the bond
                                                   of our couplehood
          to create the golden ring
                                        of the family   community
          tied with the heart's desire
                                        reaching out to pull
                                                     All of them in.
     One belly of the pregnant world
                                        giving the peaceful warmth of the stove
          to all who open their eyes.
And in the name of this Love
                    we lift our wings
                                        into the faithful wind
               of the mountain top.
     We view this wide green-blue planet

                    singing hallelujah,   singing glory freedom songs
crying crashing waves
singing low lullabyes
crying piercing pain of birth
crying gentle river of death
blazing splattered colors of sunset   clouds
                    silently turn to cold night breeze
     as we search for a nest
                         one more time together
     and pray for a rainbow   around the moon.

Grasshopper Bob

SAMUEL SCHOONOVER
Men of Honor ♂ *SCMC*

ORIGINAL COLLAGE BY
THE AUTHOR

As grasshoppers go
You are always in the flow
You are bigger than life always being the great host
And now it is our turn to give you a toast

With fingers like X-rays we are never in any fear
In fact, you're so good Dr. Gonstead himself would be
    treated here
Those vertebrae will be aligned without even a sigh
Even the hard core patients leave healed without even
    blinking an eye

As soon as we are out of the "famous chair" or up from
    the table
Faster than a grasshopper you jump in with some
    funny outrageous fable
You brighten up our spirits with a burst of good cheer
Better than any good six pack could do in a year

So when it comes to human beings you are far
    from a sinner
After making "MVP" you go on to assist all your
    friends and team-mates to be winners
You are always right up there on the top ten list
Ready to meet any challenge poised with arm up and
    hand in a fist

So we are all anxiously waiting for you to get back to
    your hopping
Because until then we will never finish our shopping
For a friend who carries such a ubiquitous omnipresent
    bright light
That jumps out at you whenever we are in your sight

From all your fans, friends, fun lovers, fanciful Bob lovers

Three Things

DEREK ALBERTSEN
Sperm Whales ♂ SCMC

Three things I have seen today:
    A bridge without support spanning a river
without water.
    A house without a chimney.
    A church without a steeple.

Three things come back to me from the night:
    A door without a latch.
    Violence without reason.
    Passion without love.

Three things to observe:
    The stream of life blowing over the trees.
    The sun's glimmer in shattered glass.
    Flowers growing in a rocky sea.

It's Dark

PAUL SAVASKY
Sperm Whales ♂ SCMC

It's dark down here
    here in the ashes.
Smoldering, crumbling ... only
    a hollow warmth
Life here is complicated ...
    burdensome.
I seek freedom and love.
In the corner, an ember
    flickers
If only for a breath of air ...
If only for a breath of air.

Stream and I

CHRIS ALLEN
Spartans ♂ EBNoM

As is the Stream, so am I.
Running fast in spots and slow in others.
Making turns and falls,
Rolling over rock and still moving onward
As is the Stream, so am I.

I Am Learning

Matt Sorgenfrei
Quiet Riot ♂ *SBNoM*

Men hunt what is obvious
    with nets already full.
They come home hungry and confused.

Women know
    for life grows inside their bodies.
Men leave no room.

Men see the night sky
    after they lost their path.
They fill their wounds with stars and
      dreams.

Thank you for the mistakes
    Which empty our traps
So we may capture something new.

Thank you for the pains
    Which force down the shield
So we may open our arms to receive.

Thank you for the tears
    Which wash away our sight
So we may see the world for the first
      time again.

Brit

John Renish
AaaRrrrr! ♂ *SBNoM*

"D'you fancy a sweet?" he asks,
Meaning, "Would you like dessert?"
    His alumINium pen peeks out
    Of his hideously checquered shirt.

He howks 'is sneck (picks at his nose),
    A truly noble beak,
And tells you of his holiday
    That lasted just a week.

"We won the wahw!" he says, and means
    The one against the Huns
With only a little late-come help
    From Lend-Lease Yankee guns.

"You have no history, no time
    To have any tradition;
You've never had an empire,
    No monarch-driven mission."

He sips his Sid (a bathtub gin),
    Gets staggeringly maudlin,
Then slumps into a fetal pose
    Till morning's heat has risen.

On Team

BILL HORST

Jerks Out Of Hand ♂ *SBNoM*

It just feels right
    being with these guys
We may not always like
    each other or agree...
and yet we are
        connected
    by a Weekend that
        bonded
    and commitments that
honor our lives
    — our uniqueness
    — our heritage

Sometimes the task or
    meeting or outing is
        just an excuse
    to get together again
Just another chance to
    be men with men —
    jerks with Jerks —
to work and talk and play
    from the gut and
        from the heart
    and to bask in our
        bond of brotherhood

Jerks Out Of Hand men's team on anniversary outing in 1991,
helping a friend fix a footbridge.
(Clockwise: Horst, Wickes, Murley, Hyatt, Landrie, Shannon,
Vlakancic, friend Brad, Ferguson)

"Alas! poor Yorick. I knew him
... a fellow of infinite jest, of
most excellent fancy."

—Shakespeare

Epitaph:

On June 12, 1993, Brad Beebe (in picture at left) was killed in the crash of the new ultra-light plane he was flying — following his passion. Brad was a close friend of Gregg Ferguson, ardent mushroom hunter, physician, and one hell-of-a-good man who touched many lives! He will be missed.

ORIGINAL PHOTOGRAPHS BY
THE AUTHOR

Home

TIM SHORT
Phoenix ♂ SBNoM

Before I was old enough for school
I found that by draping a blanket
over a desk in my room, I could build a shelter.
And from the dark inside
I could see light.
My fort would stand strong
through any storm,
a fire burned within.
Now
I cannot find comfort in such a simple shelter,
for new storms
have come into my room,
and brought with them a bitter biting cold.

ORIGINAL PHOTOGRAPH BY
MOE RUBENZAHL
Phoenix ♂ SBNoM

MEN BEING MEN

METAMORPHOSIS

METAMORPHOSIS

METAMORPHOSIS

The Way of the Sword

JOHN ROBERT BEARDSLEY
Guest ♂ SCMC

And it shall come to pass,
that all who drink freely from the
sword
will swallow an abundance of
strength, forte, integrity, and inner
wisdom.
Drawing power, and sharpening their
temper
from the magic of the sword.
Swordsmen, dancers of music, and
storytellers of history!
Dance, cut and slice, knowledge from
the past,
drink in great battles, toast bitter
duels,
and inhale torrid love affairs.
Instruct us in the passion of light!
All who raise the sword will be
challenged to the hilt
for their sweat, toil, and bondage of the
blade.
To gain their own conquest,
to free their inner expression
and awaken from their drunken stupor.

Break from the chains
Your quest begins now
Raise your hilts to the gods!

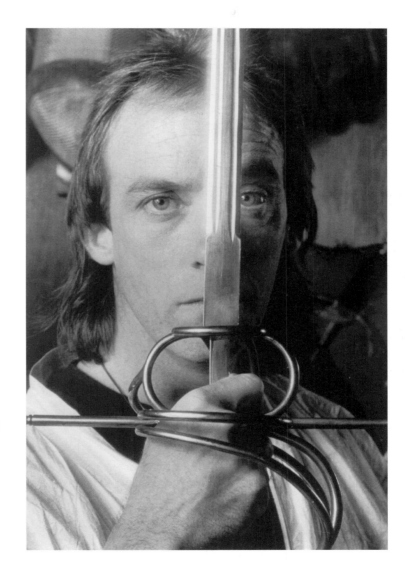

ORIGINAL PHOTOGRAPH BY
THE AUTHOR

A Question of Valor?

JOE KURLINSKI
Spartans ♂ *EBNoM*

Fog drifting lazily across the street? 0300 hours on a crisp autumn night — not right! No fog all evening! I stop my black-and-white and back it down the gentle sloping residential street to the "fog's" point of origin. As I leave the car I smell smoke, and I scan the rooftops for chimneys that might be producing the smoke. As I look at one home I notice smoke pouring from under the eaves of the roof!

As I run down the hill, heart pounding, legs pumping, automatic pilot sets in. Dispatch is contacted, a Code Three fire alarm is requested, and an ambulance is also requested. By this time I am running up the steps of the residence.

I knock (with my night stick) and yell several times to gain entry. My heart is pounding more rapidly now, my mind is reeling. No response! I attempt to kick the door open. It's solid-core and very secure. This is Oakland, you know. My best friend, partner, mentor, and (most important while on duty) *back-up*, is running up the steps. We're always in each other's general proximity. About three minutes has passed, and it seems like an hour.

"Let's kick this fucking door!" We kick together. Nothing! We kick again, and the door, jamb and all, flies into the house. The smoke is dense and about knee-high. I run to the left, he runs to the right. As I enter the room, the heat is intense and the smoke is thick. The wall of flames drives me back to the front door. Sirens are screaming in the distance.

My partner is also at the front door, driven back by the smoke and heat. We surmise that the bedrooms are located to the right and we both head in this direction. We glimpse an arm through the smoke — but the smoke and heat drive us back outside again. The sirens are screaming a lot closer.

We crawl into the bedroom and attempt to pull on the arm we located. As we do so, a fireman lifts the body from our grasp and exits the house, and we exit also.

The fireman lays the body of a teenage female on the lawn; I detect no pulse and no breathing. CPR is started, and she is revived with very little effort.

An infant and two adolescent children are now on the ground. All three are not breathing, and there is no pulse. My partner and I administer CPR to the infant as the firemen and paramedics attend to the others. The infant and a boy of about four are in the worst shape, and they are transported to Children's Hospital in one ambulance. My partner and I assist the medics with CPR in the other ambulance transporting the two young adolescents to Doctor's Hospital.

The hospital is understaffed, so we assist the doctors and medics at the hospital. We rotate the CPR and the young boy is revived after about two hours of grueling work. With the children out of danger, the doctors turn their attention to my partner and me. Tests are run, and we spend a couple of nights in the hospital for smoke inhalation. We are informed that the infant died, but the other three survived.

The newspaper crews come and snap our photos. When we are released from the hospital the television station contacts us and requests interviews. After being advised that we are too weak to come to their location, they decide that our story is not that important and decline to interview us.

I shake the Chief's hand as he hands me the Medal of Valor, the highest award issued to members of the Oakland Police Department. It seems somewhat ironic to be receiving this award, as the desire to be a police officer had been with me for the last 25 of my 30 years of life. Actions of the type done by my partner and me occur on a weekly basis with no recognition; they occur with every officer in this nation.

The realization that three human beings are alive because of our actions is reward enough. The *valor* of our actions seems minimal compared to the true reward I received that night, a night that forever changed me: I realized just how precious life is. Breathing life into a person, and assisting in the circulation of blood of another, especially a young child, makes you realize how important we are to one another. The fact that the other person responds deepens this realization.

Is this a question of valor ... or a question of humanity?

Celebrate Life

STEVE CROSS
Juggernaut ♂ SBNoM

Celebrate life
Dance the dance
Share thoughts and dreams
Tell our stories
Smoke cigars
Share life

Celebrate life
Wail out loud
Arms around shoulders
Leaning like trees

Celebrate life
Laugh at jokes
Scream like coyote
Silly cheers

Celebrate life
Men together
Count on it
Juggernaut

ORIGINAL WATERCOLOR BY
JAMES HEIDT
Bucks ♂ SFT

Stairway to Heaven

PETE DANAN
Bushwackers ♂ EBNoM

I think that's what they call it, anyway — that oversized "rope ladder" made of steel cables and eight-inch logs. As I stood looking at it my thoughts went to how much I avoid hot, sweaty, dirty, maximum-effort stuff that usually results in bruises, scrapes, strained muscles and generally includes thoughts like, "Why did I need to go and do this?"

So here I was at a ropes course on a hot August afternoon, and this guy was explaining the "rules" for climbing this ladder. My initial hopes of using the side vertical cables to get up were quickly dashed, as that was expressly forbidden. My other initial thoughts of how I would get to the top also proved premature. This was a *we* project. Two men, myself and Ter, would make the climb together.

We watched two others do it with great difficulty and extreme effort. We plotted our strategy. The log "rungs" of the ladder got further apart at every level; the top one, our goal, was about six feet above the one below it. I think our strategy finally took shape at about ten feet off the ground and was refined every five feet or so as we stood to reach for the next rung.

Looking back over those 20-30 minutes of maximum output, I recall very clearly three things. The first is the strong feeling of teamwork. No way to do this alone. No way to balance, swing up and reach for the next rung without another set of arms and legs to steady things thirty feet in the air.

The second is "no quitting." We were getting to the top this hot afternoon, or sleep halfway up for the next several nights until we did. At times on the climb there seemed no way. Strength only holds out for so long. Ideas of how to get from here to there falter and fail. Wet, slippery hands don't grip smooth logs very well. Eyes don't see clearly with stinging salty sweat flowing down into them. Breath comes shallow and rapid after a few minutes of maximum effort ... stop; rest; thirty feet above the rocks and dirt on a hot August afternoon. What the hell, let's have another go at it, Terry. "No quitting" comes back over the sound of my heart pounding in my throat. "Once more unto the breach," begins to be more than some heroic saying printed on some page read long ago. We climb on.

The third thing about that ladder I recall is the top. Up above the heat, cool breeze blowing, the view through and over the trees. Screaming with success, no longer admitting fatigue. Just elation. Me and Terry.

The bruises, scrapes, and strained muscles would heal and retreat into memory, but the two of us at the top of the ladder will be with me the rest of my days. A real stairway to heaven.

ORIGINAL PEN & INK DRAWING BY THE AUTHOR

And So It Goes …

Kevin Walsh
More Power! ♂ *SBNoM*

Some of my firefighting team and I at the aftermath of the Oakland Hills fire, 1991.

Original photograph by the Author

How to Screw Long and Deep, or
Wasting a Perfectly Good Saturday Afternoon

MARK THORPE

Phoenix ♂ *SBNoM*

There are very few tools that I would say I don't know how to use, though there are many I've never had the pleasure to hold or operate. I just figure that if I know what a tool is supposed to do, then I can make it do it. I have had a lot of success with that approach. No fear, just an inquisitive mind touched by insanity, and I'm sure I can make it work. So I buy a router, and make my own windowsill. Not bad! No one ever notices the exquisite detail of the curves I laid into that wood, but nobody mistakes it for a plank either.

I remember the first time I started playing around with the router. Whoa! Torque supreme! Just touch the trigger, and it takes off with 36,000 RPM tearing up the wood while I struggle to make it go straight. It seems that controlling circular motions is one of my recurring themes of tool mastery. I refinished a parquet floor, which if you ask anyone but me you will hear that it's not worth the trouble. You can't use a drum sander, or so I was told; rather, you need a big 16" floor polisher with a sanding disk. Some of you may have tried to use a big waxer before and have learned the technique, but controlling that monster with 150 grit on the pad was insane! In the end, though, the floor was beautiful, and my chiropractor's financial position was significantly improved.

My next door neighbor, Harry, is an engineer with all the tools I don't own and a truck to go rent the really big stuff that neither of us can convince our respective wives that we need. Harry and I have identical houses, and both of us are intent on completely personalizing them. We do a lot of the same projects, sometimes concurrently. One we talked about reached the front burner last weekend. Harry yelled over the fence, "Hey Mark, you going to

be around for awhile?" "I don't know; I've got to get out later on, but what's going on?" I replied. "I wanted to go rent that auger and drill some holes." This was something we had talked about for a long time. Our trees have been a problem in the drought because their roots keep very near the surface when there isn't much water below, so we figured we could drill holes and water them that way. "Go do it, Harry, and I'll help you drill them."

Fifteen minutes later he was back. And what a beauty! A Stihl two-stroke with a six inch bit with extension. I'd used one before at the McKinley school project about five years ago. But this stood almost as tall as me, and what a drill! We took it over to one of his problem trees. He had poured concrete and aggregate in his strip between the sidewalk and street and left about three feet of dirt on either side for the water to soak into. We positioned the auger and fired it up. Talk about torque! We were going at it, and this thing is moving us in circles as it chews up dirt and root and brings it up to the surface like an enormous ant hill. We figured the deeper the better so we let the machine chew as we struggled to keep it from spinning. Deeper and deeper.

Then we tried to pull it out. "Shit, Harry, this thing is really stuck" I said. We pulled as hard as we could but nothing. Okay, let's keep calm. "Maybe if we gas it kind of slowly as we pull up." We worked together as precision technicians. The auger dug deeper. Now it was down to the handles. "Maybe we can get it to spin until it strips out the hole, and then we can pull it right out." We gassed it, and it dug deeper still; now the handles were flared up like a lock washer gripping the curb and the sidewalk.

I won't repeat what Harry was saying right about this time, but it ended with him going inside to call the rental to find out where the fucking reverse was on the fucking drill. Minutes later we knew. No reverse. Just use the cam lock and unscrew it manually. No problem, unless you happened to bury it up to its neck right next to a tree and had the handles ground into the concrete. I was thinking, "Man, am I glad that *he* rented this," but I didn't say it. Instead I suggested running water down the screw to kind of lube things up. Good idea! If you had your choice, would you rather dig a machine out from the dirt or the mud? The mud, but of course!

By this time, we had started to get some attention. John from across the street came over. Jerry rode by on his bicycle and stopped to tell us to use the reverse. The boys in the neighborhood came by to get mud to throw at the stop sign. Everyone wondered how we were going to do it. John was simple yet elegant. "Ever see Caddyshack? We could wedge some dynamite in there and blow it out." Yeah, we could do that. My suggestion was slightly more realistic. "Harry, the problem as I see it is that the tree is in the way of our unscrewing the auger. Let's go back to the rental place and get some chain saws and a stump grinder and we can get rid of the tree!" We can also get to use a stump grinder which is one of those things I've always wanted to try out. Harry said, "This machine's design stinks. It should have a reverse and it doesn't. We can just let it dig its way into the ground until it's buried and then just forget it." Of course they would come after us, or really just him, but I felt I had to stick this one out.

My life flashed before my eyes. Children born, growing up, leaving for college. Harry and I still digging up the auger. Gas engines obsolete. Almost through. $247,000 rental. Just take his house! We opted to take it apart in the mud and using pipe wrenches and crowbars on the blade, we got it out, reassembled, and running in thirty minutes. Now that's teamwork!

The next time you notice your trees look thirsty and another root is coming up through your lawn, do what we would do: Call John Stamm.

"Killer" Chili

JOHN FREEMAN
Tatonka ♂ SBNoM

DIRECTIONS

- In a large heavy skillet, brown:
  - 2 pounds coarse ground beef
  - 1 diced yellow onion

- Drain fat and add:
  - 12 oz. bottle of your favorite American beer
  - 18 oz. can of tomato sauce
  - 1½ cups of cold water

Bring to a rolling boil. Add a substantial, or lesser, amount (depending on how "hot" you like it) of ground-up dried hot chili pepper mixture (red chilis, herbs, spices), cover and continue boiling for 10 minutes. Reduce heat to medium boil for 15 minutes until chili slightly thickens. The last stage below is **the most important**.

Reduce heat to low, salt to taste. If you like extras in your chili, add them now (e.g., pre-cooked beans, canned tomato, bell pepper, mushrooms, etc.); however, adding extras makes the chili milder. Now is the time to adjust the thickness of your chili. If you like it thinner, add a little water. Optionally, you can add 1/2 teaspoon of masa flour mixed with 1 teaspoon of cold water, mixed to a smooth paste (this thickens the brew and helps blend the flavors). Reduce heat to low and simmer for 30-45 minutes more; the longer the better. Don't forget to stir often to avoid sticking. Top with grated cheese and diced onions. Serve with hot tortillas and ice cold beer.

The Spirit of the River

DON HEALY

Spartans ♂ *EBNoM*

Every man has his own path to spiritual renewal, the fountain we all seek to drink from for spiritual youth. This is received through our own channeling, where we become one with the Spirit. Ordinary life is put on hold as we pursue our bliss.

I find my bliss in Mother Nature. In the veins of her body flows life itself. The true Spirit can flow from Nature to the soul of the river within us.

I so enjoy the river that I spend many weekends there seeking my own Spirit. Rafting is a test of the soul. I enjoy the thrill, the adrenaline rush of each rapid (it's better than drugs!), and it becomes a need that I have to fill.

I put my life in the hands of the river, and it can give life as well as take it, so easily and swiftly that there is no fighting its power. You literally have to go with the flow, and hope that the river gods smile on you today.

The river gods decided to test my faith one day on a river that I know well. Rivers can take you any time and anywhere they want. You always have to be on your guard and at the ready, every moment. This river took me on a ride on a day that's forever etched in my memory. Now, a ride can be short or it can be life-threatening. This was one of the latter.

The rapid called Quarter Mile is a fairly fast section of the river, about (of course) a quarter-mile long. But it seems like a lifetime. It's full of large waves and big rocks, with holes on the opposite sides of the rocks. In a raft it's a good challenge. But in ice-cold water with your head at bobber level it's a very different picture.

In the rapids your adrenaline is running at 200 percent. Breathing becomes a real challenge, possible only in the troughs of the waves, because you go through the crests under water. Meanwhile you have to be checking your direction, keeping your feet pointed downstream to push off rocks, and most of all relaxing and letting the river take you on her journey, putting your head back and keeping your feet up in front of you.

Toward the end of Quarter Mile Rapid there is this big hole that you can't miss, and I was getting very tired. When I hit the hole it decided to keep me. The more I struggled, the longer I stayed. When I finally gave up and relaxed, I went under, and the river spit me out again 20 yards downstream, heading for a rock and a calm eddy where my boat was waiting. I swam over to the boat and my friends pulled me in. I shook for half an hour.

I guided the rest of the trip and went on home that night, and not until the next day did my body truly tell me how it felt. I went into a mild shock for two days. There were several deep bruises that never showed up visibly but hurt much worse than surface bruises. Never before or since have I felt such trauma to my system.

The river gods were playing with me that day. They took my life and then gave it back to me all in a matter of a few moments. This left me with a healthy respect for the river — and for life itself.

I feel there are many metaphors in this story that pertain to my everyday life. Being blessed to have seen and experienced what I did has led to a spiritual fulfillment that I never would have found otherwise.

The Bishops of Bountiful

NOLAN MECHAM
Spartans ♂ EBNoM

So locked in.
 No change, no way.
  Set in script and scripture,
   to live and die by numbers.
Dream of change,
 dream of fortune,
  dream of romance,
   dream of fame.
Ruthless raging, engaged
 in perpetual deadlock with ...
  inertia.
Stillness rages in impotence.
  Importance!
  No, it's impotence.
The die is cast is die is cast. Damned
 to eternity in this iron, cast-iron
  die is cast steel cocoon.
   Maybe next time I'll be...
Damned to eternity! in this self-made
 self-pitying goddamned American Steel
  form-fitting maidenform iron maiden.
And the beat goes on, the beat goes on.
 And Sonny is the Mayor of Palm Springs!
  And Cher is the Mare of Tattoos and Lace,
   stirring your darkest teenage perversions.
Whose maidenform iron maiden is this, anyway?
 And why am I forced to....
  Who says I'm a "B" cup when I know
  I'm a triple "E" motherlovin'
   Arnold Schwartzenegger
    American Steel-pumping potent
     portant
      pinionated
      poignant
       pungent Son of a Bitch,
   at least in my dreams of fame and
    fortune and
     romance and ...
      change.

And the beat goes on,
 and the beat brazenly brandishes
  bawdy gaudy gauzy dreams of
   romance and change and fortune
   and fame.

The Mare of Tattoos and Lace
 has her way with me in my
  iron-maidenform
   and the shrapnel blows the Bishop
   and his gang of
    "Elders" and "High Priests"
back to Bountiful.
 The blood of the "Righteous" and
  blood-soaked bra bits everywhere
   testify that this is no dream.
   "Sin" sits in "Virtue's" seat
    and the whole world is
     topsy tipsy turvy
      Tinsel-Town
      wonderful.

When I Was A Boy

RAND FANSHIER
Victory ♂ SBNoM

When I was a boy, I spent a great deal of time trying to do nothing, avoiding my mother's chores, and my school's homework. Gradually, as a result of months spent lying in the grass staring up at the clouds, I became aware that I was meant to do something, so my soul-searching began. What could I spend my life doing, and not get it done too quickly? What would be a challenge, and also something nobody had ever done?

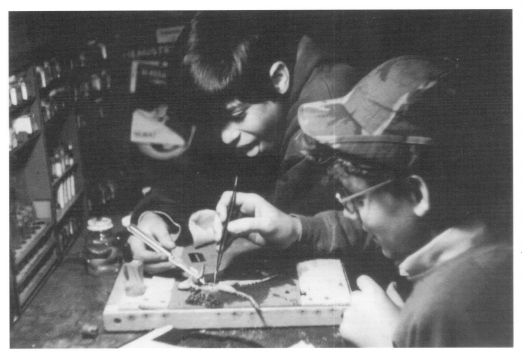

At first, at a loss to come up with anything original, I decided to be a world-class engineer, like my father. But this seemed somehow empty and hollow. With all the pressures to hurry up and **choose**, I might have stayed with that first choice, and lived a life of it, busy and purposeful.

ORIGINAL PHOTOGRAPH BY
MOE RUBENZAHL
Phoenix ♂ SBNoM

Then I got a "phone call." It didn't happen all at once, but started with some ideas that grew and grew. The hate of having to clean up the environment, of compact cars and shoe box homes, of people stifling themselves to get along with a neighbor. Then I knew. If I did nothing else, I would open up the frontier of space. I would send out probes to distant stars, in search of habitable planets. I personally would make sure that a space transportation empire was built to support this undertaking. And, if I was lucky, I would even get to take a ride myself.

With this certainty underfoot, I tried testing my conviction. I asked myself how far would I go. I discovered I could die, and I could kill. My vision was unstoppable. I would allow nothing to get in the way. Perversely, I sought barriers, just for the fun of using my pent-up rage to conquer them. With delight, I found my power growing as my energies marshaled themselves in a focused way.

For a dozen years I have been living from my commitment, and I believe that every man needs a commitment as pure and vital as mine. This is how life is meant to be lived. And how a man is meant to be used.

Happiness Is

DAVE WHITMORE, EMIL GLUEKLER, RAMJI DIGUMARTHI,
MIKE RHODES, DAVE LEONARD, PHILIPPE LAVIE
Happy Men ♂ SBNoM

Happiness is ... sipping the suds with my Men's Team on a beach in Fiji after a hard day of snorkeling.

Happiness is ... sustaining the flame.

Happiness is ... having a wife that's fun to be with, cooks, cleans, plants flowers, and takes care of herself, after an argument says "I'm sorry for being so grouchy," and appreciates me for who I am and what I do.

Happiness is ... seeing the world move and the possibilities grow in response to staying in action.

Happiness is ... lunch at the Brass Rail.

Happiness is ... jogging with my golden retriever.

Happiness is ... to be called Daddy; to see in a child's eyes the sparkles of life and joy; to come home to a loving wife; to work in harmony and comfort with my partner for life; to have men around to support and be supported by; to be professionally successful, proud of my work and financially enriched by it; to be in good health and prepared to face our world's challenges; to have constantly learned and grown; to have made a difference in this world; to see when I awake in the morning the sun rise and when I go to sleep the moon and the stars shine.

Happiness is ... watching the University of Kentucky win the NCAA basketball tournament.

Happiness is ... sitting on a ridge overlooking Royal Gorge, watching the American River work its way to the sea, and realizing that day to day problems are only manmade and how insignificant they seem compared to nature.

Happiness is ... the acknowledgment of achievement (how small or great) of goal or work, that generated satisfaction. This transformation validates oneself and people around.

Happiness is ... the four "**F**'s": **F**amily, **F**riends, **F**ishing, and giving a good **F**uck.

> *Nobility is synonymous with a life of efforts, ever set on excelling oneself, in passing beyond what one is to what one sets up as a duty and obligation. Cease in this effort and so will the laughter and tears, the happiness and joy.*
>
> —José Ortega y Gasset

Happiness and Transformation

DENNIS SIDES
Happy Men ♂ SBNoM

To my mind, much of this business of personal transformation happens in a quiet way, day by day, and has everything to do with happiness. I don't know why it should work that way, but I know it when I feel it — when I'm happy I'm operating from a higher state, and that has a profound impact on both me and my interaction with the world.

The dictionary defines happiness with words like "enjoyment," "pleasure," and "a well-adapted state of being," etc. Fine and good. To me, happiness is not the source but the result, a natural outflow from another activity — gratitude. When I am grateful for the wondrous stuff of life — a great song on the radio, good health, a day off, my team, a dazzling sunrise — I become happy. Cause and effect, action and reaction. It's that simple.

Gratitude is a participatory thing — it's more about "doing" than "being," and some days I really have to work at it. It's always worth it. When I remember to take the time and make the effort to be grateful, one day at a time, I'm a Happy Man. Life may be too short to smoke bad cigars, but today is definitely far too long a time to be without happiness.

How to Choose a Good Cigar

Todd Einck
Modern Warriors ♂ *EBNoM*

The cigar is a fascinating international pastime for many men — and a few hard-core women. The cigar is finding more popularity today than it has in past years. More people are discovering the tranquil pleasure of fine cigars as they give up the socially unacceptable habit of cigarette smoking.

The cigar was an important part of the Men's Weekend, because it represented the last vice that is truly for men. Women are not attracted to the bulkiness and robust aroma that a fine cigar produces. A carefully selected cigar will give a smooth taste and a mild aroma. Choosing a cigar is similar to choosing a fine wine. There are many cheap, poorly constructed cigars available today. If you have ever experienced a harsh, uneven burning cigar, you will know what I mean.

What is the single most important factor in determining the quality of a fine, all natural tobacco, long-filled, handmade cigar? Consistency! There are two general elements involved in making up a fine, handmade cigar: quality tobacco and quality construction. Since consistency is an integral part of a quality cigar, reason dictates you cannot judge the quality of any brand by sampling only one cigar. A box of 25 is generally needed to adequately judge the consistency. There are numerous other attributes important to selecting a cigar of top quality: even burning, ring size, fermentation of tobacco prior to construction, draw, wrapper, tobacco source and so on.

As with fine wine, it is important to store cigars in a proper fashion. If you keep your cigars at less than 70% relative humidity they will dry out over a long period of time. To combat this, get a decent humidor, keep the cigars at room temperature, charge the humidifier in the humidor with mois-

ture and you'll have no problem. The perfect storage for cigars is 73% humidity at room temperature; this will allow you to keep cigars forever.

The size of a cigar will have a dramatic effect on the taste and aroma of the same cigar by the same manufacturer. The larger the ring size the more smoke you will generate. The length is also important, but will have less bearing on these factors than the ring size. Find the size of cigar that fits your taste and stick with it. 1 ring = 1/64" in diameter; thus a 50 ring cigar is a bit more than ¾" in diameter. Cigars come in ring sizes from 28 to 52.

Price is not always the best indicator of a good cigar. Generally the biggest element of cost of a fine cigar is the tobacco. Beware of fancy packaging and exotic wrappers; they add greatly to the cost of a cigar but add very little to the consistency and quality of the product. Individual aluminum or glass tubes, polished and beveled cedar or mahogany boxes, gift selection of 5 or 10 cigars, etc., do little to ensure a good smoke.

Fermentation of the tobacco is one of the most critical tasks a cigar manufacturer must face. This process requires patience and tender loving care. A fine wine must be fermented, aged, handled and stored properly for the most favorable results; the same philosophy holds true for cigars. The tobacco must be of high-quality stock. The raw tobacco leaves are stacked into piles, called "bulks," for fermentation. The center of these bulks develops heat that should not exceed 115-130 degrees F., otherwise the tobacco is ruined — burnt out, so to speak. Once the centers reach a desired heat the bulks are turned inside out and the heat builds up

METAMORPHOSIS

again. When the heat levels off, fermentation is complete. This could occur after four turns or eight turns, referred to as "sweats" in the trade. Over-fermentation will ruin the leaf, causing it to become "spent" and lose its flavor and aroma.

Draw and burn are factors that are directly related to the construction process. If a cigar is over-stuffed with tobacco, it will burn slowly and draw poorly. Conversely, if it is filled too loosely, the cigar will burn hot and fast. A well-constructed cigar will burn even, draw well and smoke smoothly. Licking the cigar prior to and even

during a smoke will help to cool the burn and seal the wrapper. Pay very close attention to the look and feel of a cigar before buying it. A solid, hard cigar may indicate overstuffing or dryness. A soft cigar may indicate a loose roll or under-fermented leaf.

Cigar making is an art. Choosing a good cigar is also an art. Don't be fooled by the fancy packaging, high prices or familiar brand names. Experiment with several brands, ring sizes and wrappers. Once you've smoked a cigar that is top-notch, you will be convinced it is a pastime worth investing in.

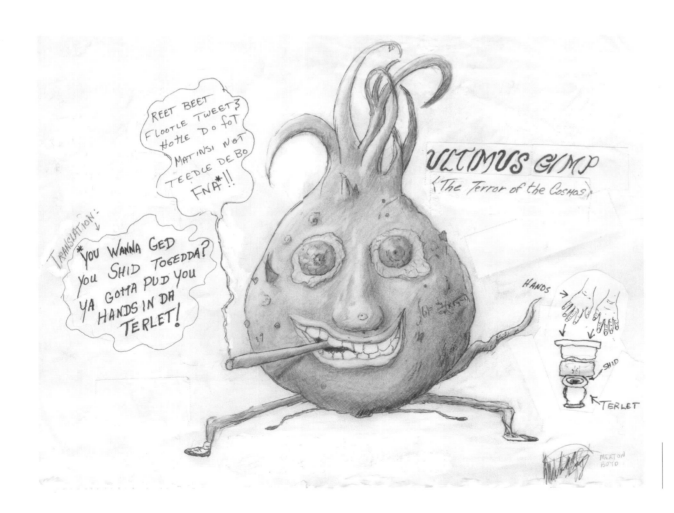

ORIGINAL PEN & INK DRAWING BY
MERTON BOYD
Centurion ♂ *SCMC*

Some Reflections on ICSD

JERRY RAITZER
NFL ♂ SBNoM

*We walked the path of teamwork,
and discovered how simple it is to
move a mountain ...*

*We walked the path of learning,
and discovered skills we had
forgotten, a new respect for our
teachers ...*

*We walked the path of commitment,
and discovered how we can help each
other to overcome our barriers ...*

*We walked the path together, with commitment
and integrity, and we changed some people's lives*

*Discover for yourself how ordinary people
can achieve extraordinary things, for ourselves,
for our children, and for our community ...*

Walk the path

*International Community Service Day
May, 1991*

ORIGINAL PHOTOGRAPHS BY
THE AUTHOR

METAMORPHOSIS

FLIGHTS OF FANTASY

METAMORPHOSIS

Olympus Mons

PAUL BOERGER
Task 3+ ♂ *SCMC*

Delaney's dead. He slipped away quietly, sometime during the night. His suit integrity had been good, but he'd been lethargic during the day and had fallen behind. He'd stumbled into the ecodome late and hadn't wanted anything to eat. The next morning he was cold and still. I'll do with him like I did the others, strip him from his suit and leave his naked body to the winds.

We'd been told about the winds, but words can't begin to describe the howling, shrieking force, that sweeps this godforsaken place. They blow at hundreds of miles per hour, storming and raging for weeks on end, moaning, screaming through the freezing days and nights. The sound fills your head until you begin to hear them speak to you. A thousand voices speaking at once, saying everything and nothing, taking the place of all your senses. And when they stop it's like a silence that could still your heart. Like Delaney's heart. A stillness that lasts forever. The winds will take care of him and in a day he will be dust. Taken by the storm, to circle this planet, forever a part of the red howling dust of Mars.

Delaney's gone, Smith and Jacobs too. I am alone on Olympus Mons. Olympus Mons! Beyond reason, this gigantic mountain vaults above the plains of Mars. It rises five times Everest and would blot out the state of Texas at its base. Jacobs used to say that when we reached the top we could wave to Earth and be seen. We'd laugh. No one's laughing now. I am the last of us and there is no humor left in me.

It was Smith's idea really, but my own secret desire danced around the cradle of imagination, and I could not let the reality of such an undertaking pass. Yes, I said, Yes! I pushed, I prodded, I would not give it rest. It consumed me and the others wondered for my sanity. I was possessed with the mountain and its majesty. Like the winds of Mars it filled my consciousness until it blotted out all else.

The space shuttle had been turned years before into a true transportation system. After the failures of the 80's and 90's it finally became mission capable. It was offered commercially as a vehicle to anyone who could afford it, and space became a playground for the millions. Travel companies refined it and payloads of adventure seekers lifted off from the Earth to destinations only dreamed of by our grandfathers.

Grandfather could tell a tale, I can tell you. My friends and I would gather around him under the still, fading blue skies of Chamonix for one of his amazing stories of the mountains. Though at the time I had terrible doubts, I found out years later that all of Grandfather's tales were true. I never thought of him as old, but my mother had told me he was over one hundred years old and wouldn't be long for this world. I never believed he would die. Does anyone, really? But die he did when I was nineteen, and as it is with the young it was years later until I missed him.

With the wisdom and bitterness of the adult world you came to understand the innocence of the relationship you had with your grandfather. He would never harm you or tell you a lie. You might not always believe him, but you knew he was only trying to make you feel smart or warm inside by telling the tales he did. But like I said, I found out

ORIGINAL PHOTOGRAPH BY
RONN LETTERMAN
Sperm Whales ♂ *SCMC*

that Grandpa's tales were true. Grandfather had been a guide all his life. He used to say his blood ran with rock and ice. He was of the old school and decried the changes that had come to the mountains since his youth. I remember him being very angry with me when we did a one day ascent of a peak that had always been done in three. "You didn't stay at the hut," he said? "You didn't really do it. If you don't stay at the hut then you have lost a piece of the mountain." Come on, Grandpa we would say, and then rush off to the next adventure, our youth racing past us like the stones beneath our feet. I didn't know it then but a lot more of Grandfather lived in me than I knew.

The company we hired to set us down had provided all the normal equipment. We were fitted in armored anti-gray suits tuned to our weight and complete with oxygen and liquid nutrient. There were camps set up along the way with fully stocked ecodomes. Navigation beacons insured we would find the route. The mountain had been done before but we had other plans.

The pilot of the lighter that set us down thought what we were doing was insanely funny. Climb the Mons! What a fucking riot. He was greeted by stony determined faces and proceeded to ignore us and to blow us out the drop tubes without another word.

It's **all** been done, it's been said; it can't go any further. 5.22b/c seems to be the limit, but how about that route going without the protection in place. Everest naked? El Cap, Half Dome, Washington's Column, Sentinel, and Astroman in a day? And without aid? It's **all** been done.

At ten thousand feet the winds drift you around and the suits compensate for themselves. By the time you see the ground you are working the retro-rockets on your back just to stay alive. We blew out the tubes and worked our way to the ground.

We all made it down. Mars is not exactly Yosemite Valley you know. The Martian version of Camp Four is the lighter that dropped us from orbit, and the pilot thought we were bunch of fucking idiots anyway. We missed the drop zone by only 20 miles (a short hop with grav suits and a near bull's-eye by drop standards) and quickly made our way to the ecodomes of base camp. Over dinner we slapped hands like old pros at this sort of thing but our eyes, which could not hold a steady gaze, told of our fears.

When Grandpa arrived in the village one morning with no pack, rope, or client, the tale began to grow. His beard was frozen solid and he was blind from the snow. His voice was strong though and he pointed to the unclimbed ridge with his ice ax and said "He is alive up there." Later his client said that he was sure he was being left to his death by Grandfather. Between the avalanches, storms of sleet and snow, ropes iced to brittleness, and the uncertainty of the route, he was sure he had been abandoned. He had cursed the name of my grandfather until his voice could speak no more. He had given himself up for dead. When he woke in the chalet near the base of the mountain and the first person he saw was Grandfather, he praised his name and thanked god for such a guide who could have braved such weather and given up the last of the fuel and clothing and gone on for help. All Grandpa would say is, "It was hard but I made it."

It was hard but I made it. It was hard but I made it. It was hard but I convinced them to shed all but the most sensible accessories and go for Olympus Mons free. Free Olympus Mons. Oxygen of course; this was not a question of how far or fast you could go without it, if you didn't have it you died in 60 seconds. Armored suits, of course. Cerro Torre was a whisper compared to the tornadoes that passed for casual breezes here. That back on Earth you said the wind blew, you covered up, but here you

METAMORPHOSIS

had best be underground or have two inches of steel between you and the air.

Anti-grav? We calibrated it up. That is to say, against all the odds, we made ourselves heavier. Instead of being able to power jump past any place we chose, we had to climb it. We were not only neutral to Mars gravity but we were Earth equal. Delaney wore an old tee-shirt under his suit that said, Fight Gravity. Smith loved it, Jacobs thought it was stupid, but only I understood what it really meant.

Delaney and I had climbed on Earth together for many years. He was the kind of climbing partner who not only gave his best in the mountains, but who shared with you his life. We talked of our lovers, our wives and our children. Although we always strove to make a higher number, the real reason we went out together was that we could share anything with each other. We understood the absolute trust that the person on the sharp end of the rope had to have in the person who held the belay. Sometimes, no one else would do, and in fact we came to actually depend on each other. I found I couldn't wait to get climbing because of all I needed to say to him. He would say to me, god it's good to see you and guess what happened between me and my woman. For most people it's hard to imagine there is anyone in the world who under the absolute worst of conditions (You're in a life boat and your child lives if you throw yourself overboard to circling sharks. The sharks have already devoured before your eyes the remains of those who drowned. The choice is clear: you or your child. You instantly throw yourself into the waiting mouth of a shark.) would give you or your children the same consideration

they give for their own. Delaney was such a man and I mourn his passing.

I am alone on Olympus Mons. I have only 20,000 feet to go. (That's as good as the South Col on Everest, you know.) I tune my anti-grav to true Earth weight. It feels incredibly heavy after so much time on Mars calibration, but my feet step in exultation, the cadence of the winds now a rhythm, giving beat to the infinite dance I play out on the barren slopes of the biggest mountain in the known universe.

We had agreed to do it straight, and so alone I keep the promise I made. Earth weight, with ancient tools. I carry in my hand my grandfather's axe. Not even an ice tool by modern standards, but sufficient to keep me going on the endless, broken slope of the mountain.

When I reach the summit I will thrust it down deep. And as the stars spiral around me and comets streak across the sky, I shall remember what brought me here. I shall not mourn anymore, for there is no longer a reason for mourning. When I crest the final ridge their spirits will soar about me.

Grandfather beckons to me and I smile with infinite peace. He tells me I have done well.

"Why are you here?" I ask.

"Time for that later, my boy. Come up to the top and get some rest."

As the last of my oxygen bleeds through the lines, I shove his axe into the summit and hear him ask me, "Now wasn't it better that you stayed at the hut, just for a night?"

Original photograph by
Ronn Letterman
Sperm Whales ♂ SCMC

Father And Son

CRAIG W. CONGDON
AaaRrrrr! ♂ *SBNoM*

Mr. MacRae was intent, proudly regarding the little robot quiet on the floor. Motionless, lifeless, it did nothing more than occupy the volume of an ordinary house cat. More like a footstool than a cat, its four identical legs joined to a thick trunk. There was no head to speak of, nor a tail. One glass eye was the remains of a dissected camcorder that swiveled horizontally on a motorized platform. The neck sprouted a neatly wrapped ponytail of wires connecting an umbilical that trailed nearly half the length of the garage and terminated at a cooler-sized beige box sporting a keyboard and a video display. The legs resembled those of the animal that the machine tried to imitate — haunches connecting to shanks, connecting to extended ankles, touching the cold smooth concrete with small rubber pads.

There was no skin, no fur and little attempt at mimicking the appealing niceties of nature's original design. There was only a skeleton of metal beams and plastic joints, hydraulic piston muscles glistening with an oily sheen, miniature rheostats, fine wire cables rounding watch type pulleys and delicate pressure sensors under the rubber foot pads. A bank of servo valves crowded the belly, and strapped to the back was a small stainless steel tank of thermos bottle size for accumulating hydraulic fluid — blood for a synthetic organism. Every square inch bulged with components, colorful wires, plastic and metallic tubing, metallic this's and metallic that's, a veritable smorgasbord of widgets and gadgets all bundled carefully to function as a small quadruped. Each component, nothing more than inanimate trash, suddenly springs to life in the hands of the technological magician when joined together in just the right way.

The old man's heart pounded with the anxiety of anticipation that he had always experienced at this, the moment of truth. To him the "smoke test" was aptly named. Engaging the "on" button was always a curious blend of fear and joy. He recalled the time sixty-one years past when an inquisitive seven year-old tested the electrical properties of a small electrolytic capacitor by plugging it directly into a wall socket. The little component ended its simple life with a flash of explosive light and smoke just inches from the young boy's eager eyes. Startled reflexes thrust the lad back skidding on the floor like the touchdown of a ballistic newspaper impacting a hard driveway. A memory, forever etched in his brain, ready to emerge whenever the electrons of life beckoned his call. Foolish though he knows, he will face the minor terror once again. A nod and an inward smile will urge him on. For on this day, complete at last, his little creation is ready to walk for the first time after more than ten years of passionate devotion. A decade equivalent of evenings, holidays, weekends and vacations invested in a synthetic creature that won't even purr, lick his fingers or piss on the rug.

Wife dead for a decade and estranged children thinking him an old fool who should be put away, MacRae works to fill the void within his life with things that bring him comfort, pleasure, and satisfaction, fanning the fire of an old passion. Faith and comfort were found with the playthings of science: meters with dancing needles, gages translating properties of matter into the language of science, metal hungry machines that regurgitate works of art and precision, digital microprocessors no bigger than a postage stamp yet able to guide a missile to a target, operational amplifiers, such and more. His toy box was ever growing, never quite full enough.

METAMORPHOSIS

Still transfixed by his creation, the aging tinkerer reached automatically for a battered old pipe laying hidden on the bench where it disappeared within an alluvial fan of soldering irons, volt meters, PC boards sprouting wire hairs and a seemingly infinite host of other bric-a-brac of his play. The pipe thrust with a smack of hairy lips into his open mouth, the mouthpiece cracked and caked with moist tar. Its comforting fire had extinguished moments before. Discouraged and too impatient to re-light, he placed it back on the bench top where it once again vanished in the clutter.

There was no more waiting. The moment was at hand. Each step had been painstakingly and meticulously thought through. Each subsystem had been tested individually and successfully. Now was the moment when each of the solitary parts would be brought together and play as one. Now was the moment that he had been waiting for. He fingered the power switch on the large box, applied a sharp vertical thrust, then jerked his hand away.

The switch snapped like a starting gun. Inside, a cooling fan whirred to life. A power supply began to energize circuits which pulsed with a team of many dozen parallel processors, a billion transistors, a billion ways to say on and off, one and zero, yes and no. The display winked into life with an audible click and a scrolling of text and numerals.

Abruptly, the small mechanical beast on the floor stiffened and straightened its legs to full extension. The man shot his hand to the power switch and froze. Uncertain, cautious, protective of his creation, he was ready in a heart beat to cut the power in the event something went awry. The old man was motionless. Laser beam vision picked apart each action of the little machine with analytical detail. The machine just vibrated, a frame locked on four stiffened columns.

Seconds passed like millennia. The legs slowly folded in unison until the belly was barely two

centimeters from the floor. Only the cyclopean eye swiveled back and forth redundantly scanning a narrow field of view directly in front of itself. The man dropped his hands to his knees stooping and staring with detached fascination at the creature before him. He slowly circled the little robot until he was sure that his legs appeared in the creature's field of view. The mechanical eye panned once and fixed on the motion of the walking man. Mr. MacRae could see the iris dilate and the zoom lens adjust in what he interpreted as an attempt to accommodate his entire form.

Suddenly animated with delight, the old man lowered himself prone to the floor, eyes wide with anticipation. He groaned slightly from the exertion and placed his warm palms coolly on the garage floor, never once removing his sight from the subject, even for an instant. The ambiance of the whirring fan orchestrated with a steady series of seemingly random clicks from the solenoid valves. Slowly, almost imperceptibly, the creature began to bob up and down with mechanical precision. Within a minute, MacRae's impatience surfaced, one eyebrow rising to a tall arch of furrowed skin on his forehead. The machine instantly froze. MacRae's other eyebrow shot up to match its neighbor.

The machine stopped in mid-cycle, its glass monocle fixed on MacRae. The man returned the stare with equal mechanical intensity. Nothing was happening. Nothing, to the eye that is. But MacRae knew the significance of this nothing. He knew that something very meaningful was happening. Something was going on that was beyond motions, beyond words, beyond the ticking and whirring before him. Something was happening inside. MacRae knew that the real events were transpiring within the beige box that housed the silicon brain of the creature. With a start, he pushed the floor away and rose awkwardly to his feet. He darted to the monitor and inspected the placid screen. All that was visible was the last load command for the creature's behavior algorithm. Pounding the keyboard he deftly entered a new command. The screen burst into a continuous scroll of unintelligible hexadecimal characters. "Ahh," the aggravated cry broke the mechanical drone of the fan.

MacRae's head turned sharply, eyes now round and mouth agape. He looked down his shoulder at the activity occurring near his feet. The creature straightened its legs crisply to nearly full extension. A leading appendage began to rotate rearward, sliding the foot pad smoothly on the cold floor. At the end of the stroke the leg lifted, rotated forward, extended fully, found the earth and repeated the backward slide. It resembled a bull readying for a charge. MacRae stood fascinated. Within moments the hip joints of the other legs released and allowed the torso to shift forward with the next drag of the foot. This time the foot did not slide. Instead the animated figure stumbled forward clumsily. Instantly, the opposite front leg lunged forward and checked the unstable motion. The machine had taken its first step. Like a newly born calf the mechanical creature shook awkwardly, uncertain, testing, learning of gravity and how to oppose it. Again the first leg extended and reached forward. Now, a rear leg joined the wobbly cadence, and the creature began to jerk forward.

Breathing for the old man was suddenly labored as if each full lung had to be filtered through the beating wings of a small bird in flight. He felt his larynx constrict, and his eyes grew glassy. MacRae turned to the mobile machine and knelt to the floor as if to pray. He watched his little plaything teach itself to walk. Wringing wet with emotion, a tear streaked his cheek, the humming fan padded the air in accord with random clicks of the living servo valves. The lens scanned hard to the left and fixed on the kneeling human a meter away. The animal seized mechanically, swayed slightly then reversed direction awkwardly. Tiny oiled cylinder shafts winked in and out burnished in the fluorescent lights.

MacRae longed to embrace his creation as if it were a real living pet. "It is real," he thought. "Here is my Pinocchio and I am Gippetto. The good fairy is truly good. Oh, how wonderful this is. How happy I am." He restrained himself from grabbing the little thing when he saw it beginning to turn. With each lumping shift and pull the critter pivoted slowly, executing a pregnant pirouette. In quick time the ungainly creature had the man focused directly in front of it and paused. Once again the forward motion ensued.

A jovial roar welled up deeply from within the man, extending his hands to greet his child. "Come to Pa Pa," he choked with painful ecstasy. With less than half a meter to his comforting grip the front legs rose high and pawed the air. The umbilical life line had snagged on a protrusion and seriously inhibited the critter's forward motion. Rear legs continued stride with pistons pumping. Before the clumsy beast could over-center and capsize the aged engineer thrust headlong, arresting the mishap, elbows cracking sharply on the unyielding concrete.

Heedless of any pain, he drew to a sitting position, still grasping the twitching entity. He folded his arms around the little body of metal and plastic sinew. A small amount of oil leaked onto his trousers from God knows which tube fitting. But that didn't matter. The oily spot grew and the little beast seemed to calm from the man's cradling arms, its solitary eye darting in small arcs. Peace fell warmly upon them both. One mechanical foreleg pawed gingerly at the air tapping the man on the forearm. He grabbed the leg firmly in a fist. There was no resistance. "You are alive. You are real." He wept with joy.

ɓ

Months passed, and the robot was transformed slightly with the addition of a nearly vertical torso section that sprouted two articulating limbs folded

METAMORPHOSIS

close to the creature's component-riddled body. The camera lens was perched atop the structure improved by the addition of vertical motion. Now it could scan up and down as well as side to side. Like a gigantic praying mantis it stalked its cluttered domain, probing and investigating the world within the range of a newly lengthened umbilical. The arthropod had free mobility throughout the garage. Slowly, patiently, with the curiosity and dexterity of a toddler it touched, poked, pushed, pried, and tugged at objects within the grasp of each of its three-fingered extremities. Occasionally, it would find a curio of special fascination, usually one that it could lift with the small articulated pincers, carrying its new prize away. A corner of the garage hosted a growing pile of hardware and small test equipment.

A bump and a whir gave pause to Mr. MacRae's captivation by the half-completed mechanism on the bench before him. Lifting his attention, the man's frozen countenance melted, suddenly transformed into a beam of joy and satisfaction. His gut twitched with pangs of pride dwelling momentarily on the comical insect pawing at a loose incandescent light amidst the eternal benchtop clutter. The curious critter had delicately coaxed the bulb to table's edge and past that barrier which isolated the hapless article from the planet below. A sudden pop disrupted the scratching and fidgeting, MacRae's smile unwavering. The bulb exploded into a cloud of white powder and countless fragments. A mechanical eye swiftly followed the motion and witnessed the demolition, arms outstretched helplessly. With abrupt awareness, the machine snapped motionless. The camera lens rotated from the shards on the floor to meet the gaze of the old man. MacRae issued a mirthful chuckle dropping his attention back to the metallic skeleton before him that closely resembled a complete human hand.

ᛡ

Nature requires several million years to create beings as complex as a turtle or as enigmatic as a human. Mr. MacRae needed only slightly more than a decade to evolve a giant preying mantis into a box supported by two dwarfish legs bulging with cables, cylinders and lumps of plastic gizmos. MacRae cast a satisfied eye upon the mechanized entity waddling down the hall. The beige cooler could now walk bipedal and untethered. Self contained, the machine had no need for power or control lines. The creature now possessed the intellectual capacity of something akin to a chimpanzee. Unlike a chimpanzee, it had no arms, hair or even skin. MacRae reasoned that with this model it was enough to accomplish the small miracle of walking on two feet. The ungainly creature rocked side to side with each step like the pendulum of a cuckoo clock. The familiar clicking of valves was joined by the soft drone of turbopumps and the rhythm of padded soles slapping the tile. When it was hungry it would seek the nearest receptacle and plug in for a feast.

Keeping pace with the changing times of technology MacRae was able to incorporate the latest geewhiz gadgetry into his maturing creation. Organic circuitry supplanted the mundane silicon. What was once a scrap book mosaic of digital ones and zeros now has become a masterpiece of fine art with all of the values in between. Over time, MacRae's fondness for the robot grew into more than the regard one could have for a favorite toy. He came to refer to the machine affectionately as Daren, the name of his only son who died soon after birth over one-half century ago.

MacRae hobbled slowly to the long table standing sturdy in the living room's center. A waterfall of long white hair and stringy beard flowed around a balded pale head. Flesh drooped like melted wax around tired passionate eyes. Bony fingers stretched with spotted skin reached for the edge of the stark white sheet covering a massive figure in quiet repose.

He peeled the cover revealing a man in metal armor. Its smooth contoured face was mouthless. Two glass black eyes shimmered like bottomless mirages on a rolling sea of glistening bronze. Small plastic hoods cupped each ear hole that nested in a confusion of heat sinks, cables and tubing. The thick neck tapered wide to the armor-plated torso. Segmented plates of plastic and polished sheet aluminum comprised the abdomen and back. The arms and legs were thick skeletons bulging with tubing and cylinders. Hands were delicate with four long padded fingers and opposable thumbs.

MacRae inspected the giant, beaming with satisfaction, his eyes now aloft and fired with passion. The knock of plastic feet arrived in the doorway. He looked to receive the mechanical stare of binocular camera lenses atop the two-legged box. "Soon, my little friend. Very soon." He staggered to the cabinet fetching a small torque wrench.

Holding a silver serving tray in his mechanism callused hands, Daren stood tall in the bedroom doorway, the faint hum of turbopumps entering the room with him. His glistening head turned with a faintly audible whine, glass black eyes focused lidlessly on the old man face up on the bed. Except for the mechanical sounds of pressurized fluids, active servos and the occasional metallic snap of alloy against wood, the man-shaped machine was silent. He stood motionless by the bed, tray still in hand. Tall and graceful, still riddled with plastic tubing and plastic and metallic gadgetry, the android epitomized the pinnacle of the master craftsman's life work. The oval face looked down expressionless at the aged man in repose. Placing the tray on the night table and reaching out with one mechanized hand he gently touched MacRae's shoulder. "Mr. MacRae?" came the perfectly enunciated voice with mechanical etiquette. "Mr. MacRae. It is time for your medication. Please wake up."

The old man raised his pasted eyelids. A smile appeared somewhere under the flowing gray beard, tired eyes raised with the soft glow of pride that they always did when his synthetic friend and companion entered the room. The metal man dropped to the edge of the bed simulating a sitting position. He maintained most of his incredible weight on the bed frame so as not to excessively tip the mattress and bring any discomfort to his frail master. The android pivoted at the waist, grasping the cup of water from the tray in one hand and presenting a small yellow pill in the other padded palm emulating the picture of perfection for bedside manners.

MacRae attempted to reach for the tablet but his weakened state prevailed and the arm of tired flesh dropped exhausted to the covers. "Do not exert yourself." The machine's voice was not cold. There were no lips through which words would issue. There was no flesh over the smooth metal face. Daren had little expression that was nonverbal. His voice was soothing, caring and sincerely reflected the feelings of the man-made soul inside. He re-tabled the cup and mindfully plucked the pill from his own palm, carefully placing it between the waiting lips of his flesh and blood companion. With the skill and care of a loving mother, a cold hand gently scooped the old man's head, lifting it slowly above the pillow like a precious ceramic sculpture. The old man's fine white hair nested between the soft plastic-lined fingers. The water washed coolly over MacRae's parched and swollen tongue. With care and precision his head lowered to the cratered pillow, old blue eyes fixed gratefully on the expressionless face. The inhuman hand had a more than human touch as it slowly slid from behind MacRae's head and tenderly palmed the side of his face.

MacRae's eyes rested wearily like two old gems at the bottoms of darkened pockets. Age spotted and heavily wrinkled, the flesh clung pink to the centenarian's skull. A sigh issued from his smiling mouth, fine gray whiskers fluttered in the frail breeze. The skin of his temples folded in neat little furrows and his head tipped slightly toward one shoulder. He beamed with love into the black fathomless glass lenses that stared dispassionately at him.

Daren rested his hand firmly on the drifting man's shoulder like a statue of weld art. MacRae tried to speak but suddenly winced. His eyes popped to wide ovals, fleshy furrows vanishing. With a desperate burst of strength he shot his hand high and hooked it over the mechanically muscled forearm stretched over his chest. The old man took one deep agonized breath, peacefully closed his eyelids and went to sleep for the last time. The limp hand slid lifeless to the covers. The room was silent with the muted afternoon sunlight filtered through heavily curtained windows.

The android placed the cup back on the silver tray, raised himself tall from the bed and started to leave the room. He paused before passing the doorway. Turning slightly he swiveled his head to gaze briefly with uncertainty and hesitation at the motionless figure on the bed. Daren hoped the master would greet the morning well with the night's rest. He thought it good and turned, leaving the room.

Next morning, Daren entered the room tall and graceful as he had for many weeks, carrying breakfast on the same silver tray. "Good morning. Breakfast, sir," he announced joyfully from the small crystalline speaker in his neck. The android was adept at creating cheerful conversation to boost the spirits of his bedridden companion. "And a fine morning it is too. I believe that the delphiniums are finally blooming. A hummingbird, the first one I recall seeing this year, appeared at the feeder I placed outside the kitchen window. It was beautiful. I hope that orange juice is appropriate with your oatmeal." The man on the bed did not move. The android glided to the night table

and bent to place the tray, his stare held on the quiet man's empty countenance. A metal hand gently jostled the soft cold shoulder. "Sir. Time to wake. Sir?" Daren dropped to his knees with the sound of plastic heavy on hardwood floor. His stare shifted from side to side in a hopeless effort of finding something to distract himself from the hideous truth. Mechanically he began shifting the covers to comfort his friend. A shaking metallic hand gently clamped the still, lifeless hand. "No," the robot crackled distantly.

If cold metal could feel colder, it had. Somewhere deep in the semi-organic consciousness of the metal man stirred the denial of observed reality. There was nothing in Daren's programming or experience to help him deal with this. For the first time in his life he felt alone with the stabbing pain of loss. A metal hand covered black glass eyes. His head drooped. Tearlessly, the synthetic man moaned with real grief.

ORIGINAL PHOTOGRAPH BY
MOE RUBENZAHL
Phoenix ♂ *SBNoM*

Business As Usual

GLENN DAVIS
Grunt 'n' Pokers ♂ *SBNoM*

Jim pursed his lips into the warm trickle of water and grimaced as it gurgled bitterly down his throat. He tried to remember how many glasses he'd had that day. Not enough, he determined. Not nearly enough.

He stood up straight, more or less, and silver worms swam across his field of vision. They resolved into a deep broth of twinkling dust motes, lit coldly by construction lights, in a wide drywall corridor stretching off into infinity. He sighed and started walking, straightening his tie as he went. The clopping of his feet echoed loudly around him.

He stopped in front of a huge, intricate, framed map of the Corporation, with a rainbow colored arrow that said: You Are Here. As he stood glaring at the map, he heard a sound coming from far away, swelling to familiarity; he turned and saw Old Bill leering at him from his mail cart. Bill's thin, age-beaten body hunched over the steering wheel of the cart like a giant wrinkled claw, with a weird, pointy head stuck at the wrong angle between its fingers.

"Want a *ride*?" Bill shrieked.

Jim cleared his throat and straightened his tie. "Sure," he said, smiling dimly. He walked over and sat in the padded orange seat next to the old man. The electric motor hummed to life, and they began rolling down the hall. The cart traveled very slowly, but Old Bill bent down over the steering wheel as if he were driving a motorcycle going a hundred miles an hour.

Suddenly he rotated his misshapen head to stare wildly at Jim with his good eye. "My, don't we look *pretty* ta-day Mr. Jim!" he said, in his bizarre voice.

"Where you be the taking ta-day, sir?"

Jim stared uncomprehendingly at the old man. Like many people, he lived in a perpetual haze of fear, which things others said sometimes did not penetrate. "How are you, Bill?" he asked, not because he cared, but to kill the fear.

Old Bill rocked back and forth with glee, as if Jim had said something which delighted him. Then he stopped and leaned toward Jim conspiratorially. He said: "Questions can open up the inner darkness of a man's soul. I wouldn't be asking such questions if I were you, unless you're prepared to invoke the *answer*!" He leaned back and howled with laughter, as if he had said something hysterically funny. Jim winced and shrank a little into his seat. Serves me right for asking, he thought.

"Now sir," said Old Bill, regaining control of himself, "Where you be the taking ta-day?"

Jim suddenly realized what he was being asked. "Mr. Bob's office."

"Good choice sir, very good choice. For that we'll need, the ela-vater." They turned a corner and stopped in front of the huge steel doors. Old Bill leaned out and struck the UP button. Booming echoes of machinery issued forth. Presently the doors slid open and the cart rolled inside.

The elevator was a sterile, fluorescent white chamber. Old Bill reached out a gnarled hand and punched one of the lighted buttons. Without withdrawing his arm, he pointed to an unmarked row of buttons below the GARAGE button. "Do you know where those buttons go?" he rasped.

"No."

METAMORPHOSIS

"I'll tell you where they go," Bill said in a hoarse whisper. "They go to the bottom of the ocean."

The cart rolled into a gigantic white room. The ceiling was soundproof tile and fluorescent panels, with sprinklers every ten feet. The room was awash with noise, the noise of people talking, telephones ringing, computer keys clicking, printers printing, copiers copying. It was the sound of business. The room was separated by orange partitions into hundreds of small cubicles. Bill's little mail cart whined to a stop in front of one of the cubicles, and Jim hopped out, relieved to get away from the eccentric old man. Old Bill dropped a handful of envelopes into the "in" basket on the desk of a startled secretary, and rolled away, cackling to himself, "one for *you*, aand one for *yooouu* ..."

Jim straightened his tie and walked into the cubicle of Mr. Bob. Mr. Bob's cubicle was a picture of order and perfection. A tiny palm tree stood in the corner of his cube, perfectly symmetrical between his filing cabinet and his computer terminal. Bob himself was the perfect businessman in a blue suit and a red power tie. His starched cuffs stuck out at exactly the right length from his jacket sleeves. He sat at a Formica desk, in a padded powder blue chair, staring at a framed picture that he was holding in his hand.

Jim sat down in one of the orange chairs facing Bob's desk. "Bob," he said, "I've just been down to Production Contract Scheduling Control. I've been checking up on this problem with the 220-RFA scheduling requests. I talked with Chuck about getting a requisition for a servicing route report, but he told me that I have to submit a 330 to Contract Servicing. But I can't get a 330 without clearance from Contract Administration. That's why I'm here to see you."

Bob sat still for a moment. Then he slowly raised his head and looked at Jim. "Well hello, Jim!" he said, smiling wanly. "Why don't you have a seat."

He went back to staring at the picture in his hand.

Jim frowned, perplexed. "This is serious, Bob," he said. "If I can't get clearance for a 330, I'm stuck. I'll have to go apply for a 55SD. And that could take weeks. I need your help to get this pushed through."

"That's nice, Jim," Bob said sleepily. Carefully, moving as if he were drugged, he reached over and set the picture down on his desk. It was a photograph of himself and his family. His wife and children stood in the background of the picture, frowning fiercely. He was down on all fours in the foreground, with a grimace of pain on his face. "I'm too old for that kind of thing, Jim," he said. The tone of his voice shifted from boredom and despair to dreamy wonder as he drawled, "Do you ever think about the meaning of life, Jim?"

Jim couldn't think of anything to say. He straightened his tie. "No," he said.

Bob looked at Jim with a far-away smile. "I think life is like a dream," he said, "and all the people are just people in my dream! And if I ever woke up, everyone would just disappear! Just...pop! Like a soap bubble! Do you think that would be a good thing, Jim?"

Jim stood up. "I think I'll go get you a cup of coffee, Bob. Does that sound good? A nice cup of coffee?"

Bob frowned. He seemed to remember something. "Cup ... of coffee. Coffee ... cup. Cup. Coffee cup." He reached down, opened a drawer in his desk, and pulled out a black coffee cup. He set it down on the desk and looked at it, frowning. He looked down at his lap and sighed raggedly. Then he straightened up and looked at Jim, all at once the stern businessman. "Jim, did you discuss this problem with Contract Control?"

"No, Bob, I wanted to run it by you first."

"Well, then, let's take a look at it over a cup of coffee. By God, somebody's got to do some work in this damn place."

Jim let out a long sigh of relief. Thank God, he thought. Now we can get back to business as usual.

⌀

Bull Turkson was puzzled. He woahed his horse and waved for his men to stop. He took off his ten-gallon hat, wiped the sweat off his brow, and looked around.

Just a moment ago they had been riding through the prairie toward home. The fog had been thick, and several times he had not been sure they were going in the right direction. Now he didn't know where the hell they were.

They were standing in a hallway, wide enough for three horses to pass through. They seemed to be in a building the likes of which he'd never seen. Weird steady flames in glass globes. Smooth yellow rope and chicken wire. Red seltzer bottles that sprayed soap.

He turned to look at his men. They were tired, angry, and frightened, and he knew that one wrong move could provoke violence. This had to be handled very carefully.

"Well," he said, spitting out a wad of tobacco, "looks like we're gonna be late for supper."

Little Whipper was terrified. "What's goin' *awn*, boss?!" he drawled in his thick Tennessee accent. "Where *are* we?"

"Ah don't rightly know."

The men started muttering to each other. There were seven of them: Bull, Little Whipper, Mad Dog, Swede, Buster, Jed and Buck. They had just crossed the state line into Texas after the long ride

back from a failed attempt to rob the Grand Funk Railroad. Three men had been killed. The only thing keeping the rest going was the thought of getting home to their families, sleep and supper. And they were angry. Most of them blamed Bull for the failure of the robbery although none would say so.

Bull reached into his shirt pocket and pulled out his tobacco pouch. He was almost out. He realized that, if they were lost, he might have to go without chew for several days. This was not good. Without tobacco he would very soon turn into a wild man, killing men without provocation, raping women, slaughtering animals, profaning the name of the Lord. He sighed. He did not consider himself by nature a violent man.

Suddenly Buck spoke out. "This is your fault," he said menacingly.

Bull stuck a pinch of tobacco in his lip. "What's that, Buck?" he said, without looking at him.

"You heard what I said," Buck spat. "It's your fault we're lost. It's your fault mah brother got kilt." His voice became high and shrill. His face held all of the grief, anger and terror of the past four days. "Ah think this is a trap. Ah think yore out ta double-cross us."

Bull flexed his gun hand. "Them's fightin' words, Buck," he said. Then he looked at Buck and smiled. Buck's eyes were wild with panic. His horse grew restless under him. His hand crept toward his gun. The rest of the men sat utterly still and silent.

Bull's grin widened. "Let's talk, Buck," he said in a fatherly tone. With one graceful motion, he leaped off his horse. His spurs jangled as he hit the floor. He motioned to Buck. "Come on. Let's talk."

Buck's nostrils flared. Sweat poured out from under the brim of his Stetson. He glanced quickly

METAMORPHOSIS

about at the other men, crazy with fear. They all were staring at the floor, the walls, not moving, saying nothing. Buck slowly eased himself off of his horse, keeping his eyes glued on Bull. Bull smiled gently.

Buck stood his ground as if he expected Bull to draw his gun. Instead, Bull walked easily over to Buck and put a brawny arm around his shoulders. The two walked away from the rest of the men and horses.

"Buck," Bull said, "you and me and your brother, we've always been friends, haven't we?"

Buck gulped. "Yep."

Bull pulled a palm-sized red book out of his vest pocket. "Do you read your Bible, Buck?"

"Yes, Bull." He was lying.

"Do you know what it says in here, Buck? It says you should love your friends as you love your brothers. Do you believe in the Bible, Buck?"

Buck gulped again. "Yes, Bull."

Bull put the book back in his pocket. "Bo wouldn't want you to do anything the Lord didn't want you to do, would he now, Buck?" Bo was Buck's brother.

Tears were streaming down his face. "No, Bull."

"Buck." Bull faced him and clasped his shoulders. "I just want us to be friends. We're friends, ain't we? After all we've been through?"

Buck managed a smile through his tears. "Yes, Bull."

Bull smiled his fatherly smile. "That's real good, Buck. I always like my friends to die happy."

Faster than lightning he drew his gun and shot Buck in the heart.

The sound blasted and rang weirdly through the hall. Buck fell to the floor like a sack of sand, raising a small cloud of dust. The horses were upset. A wave of shock passed through the men.

Bull holstered his gun and took the book out of his pocket. "He was a good man. Amen." He crossed himself.

He turned to face his men, who were struggling to maintain control of their horses.

"Anyone else got anything they'd like to say?" No one did.

"You're all very ... " he searched for the right word. " ... well-mannered today." With that he began chuckling, quietly at first, then louder and louder until he was practically screaming with laughter. The laughter slowly caught on until all of the men were laughing hysterically.

"Mad Dog" Mulhauley opened a door into a black void. He scratched his bald head. Need a shave, he thought. "Hey boss! I think I found somethin'!"

There was a sound of horseshoes clopping on tile. Mad Dog saw Bull round a corner and enter the corridor, followed by Jed. The two men pulled up a few yards away from Mad Dog and dismounted.

"Well what have we got here?" said Bull. He peered into the musty darkness. He could see plain enough that the room was gigantic, twice the size of MacGooligan's barn back home. That was good; the horses were getting claustrophobic. "We'll bed down here," he said. "Round up the men."

The men and horses were rounded up quickly and herded into the cavernous room. There was no wood to make a fire anywhere, but they carried one lamp. They clustered about the feeble light in the massive, gloomy darkness. They talked little.

As he lay down to sleep, Bull took the little red bible out of his pocket and clutched it to his chest. Actually it wasn't a bible at all. It was a copy of *Poor Richard's Almanac*. But that didn't make any difference to Bull. He couldn't read.

&

In the cafeteria, Jim ordered cottage cheese and poached tofu and Bob ordered seven cups of coffee. They sat at the table farthest away from the windows. Bob hated to sit next to a window. "Reminds me too much of being outdoors," he said.

Bob was slurping his coffee as he talked. "(Slurp) what you've got to do, Jim (slurp)," he said, "is go directly to the (slurp) god damn Contract Support Control Department (slurp) and get one of those damn 230 Contract Request Release forms (slurp). And if they give you any god damn trouble (slurp), well you just leave it to me (slurp)."

"So what you're saying is there's nothing to worry about?"

"Of *course* there's something to worry about!" he bellowed. "There's *always* something to worry about in this god damn business!!" People at the other tables were looking at him strangely.

"Bob," said Jim, "will you keep your voice down?"

"You gotta be *tough* in this business, Jim! Tough! TOUGH!!" He was slamming his fist on the table for emphasis. Coffee cups rattled and spilled on the formica tabletop. He picked up a half-full cup and downed it in one gulp.

Jim scooped a forkful of tofu into his mouth. He wanted to finish and get out of there as quickly as possible. When Bob was in this kind of mood there was no telling what he might do or say. "When do I need to have that form turned in, Bob?" he said around a mouthful of food.

"Well by hell, let's go get it *now*!" Bob said. "I'm sick and tired of waiting for those god damn pansies to get off their dead asses and do some *work*!!" He was raging now, out of control. He downed a fifth cup of coffee, slammed it down on the table and stood up. People at nearby tables were picking up their trays and moving elsewhere.

Jim hastily scooped up the last of his cottage cheese and stood. Bob whirled about and strode toward the exit. Jim trotted after him. The two men walked out into the brightly lit hallway and down the hall to the elevators.

Bob fiercely punched the "up" button. "We'll show those god damn clowns what's what," he snarled. The door slid open, and he and Jim stepped in.

He hit the button for floor 37. He stared at the panel for a moment, at the unnumbered buttons below the garage. "Wonder where those buttons go?" he wondered. "I never saw those damn buttons before." Jim said nothing.

The doors slid open. They were not on the 37th floor. They were on the floor Jim had been on earlier, the one that was under construction. They stood and waited for the doors to close again. After a couple of minutes Bob punched the "close door" button angrily. Nothing happened.

A blotchy pink hue rose in Bob's face. His forehead crinkled with hysterical anger. Jim prepared to duck. "WHAT'S THE MATTER WITH THIS GOD DAMN ELEVATOR??!!!" Bob exploded, striding out into the musty hallway. "When I get my hands on those FUCKING BASTARDS I'll smash their GOD DAMN ... " he stopped.

Jim followed him into the hallway and stopped short. Bob was standing in what appeared to be a large pile of fresh horse manure.

Bob slowly looked down at the pile around his feet. He lifted one leg with his hands and looked at the sole of his manure-covered shoe. "What's this on my foot?" he said quietly.

"It looks like ... horse shit, Bob," said Jim.

"Horse ... shit? *HORSIES*?" he whispered. He lowered his leg and stepped out of the pile, gazing down at it. A beatific smile spread over his face. "Horsies," he said, breathlessly.

Jim was getting worried. He didn't like the tone in Bob's voice. "Maybe we should get out of here, Bob," he said.

Bob looked up. His mouth hung open. He was grinning, staring off into space. "I used to ask my mommy and daddy for a horsie every Christmas," he said in a faraway voice. "I wanted to be a cowboy." He looked at Jim. "Do you remember those dreams we all had when we were kids, Jim?"

Jim took Bob by the arm and dragged him into the elevator. He hit the button to take them to the cafeteria. Bob needed some more coffee.

Fred stepped into the elevator, whistling "Masquerade." He was a happy man. Tomorrow he would be moving into his own office, a result of being promoted to Market Engineering Field Manager for the entire Corporation. My daddy would have been proud of me, he thought. Fred Senior had been poor, had broken his back in a steel mill so that his only son could go to college. If only daddy could see me now! He thought. A top executive in the largest company in the world! The thought of his father's pride brought tears to his eyes.

His new office was in a part of the building that was still under construction. He was on his way to look at it. He punched the button. Gravity increased for a moment as the elevator started to move.

A few floors up, the elevator stopped, the doors slid open, and Old Bill rolled in on his mail cart. "Hey, Old Bill, how you doing?" Fred beamed.

Old Bill's maniacal face seemed a bit more solemn than usual. He looked Fred in the eye and smiled a little. "Considering the state of the Universe, sir, I'm doing well. Very well. How you be, Mr. Fred?"

"I'm doing just fine, Bill. Just fine. I'm going to check out my new office!" he was grinning from ear to ear. "What are you up to?"

Old Bill looked away, sadly. "Do you happen to know just how big this building is, sir?"

Fred thought about that. "No, I can't say as I do, even considering how many years I've been here. He shrugged. "I've never even seen very much of it. It's like a maze, you can get lost. I try to stick to my own department as much as I can. Have you seen all of it?"

"No sir," he answered. "I don't know how big it 'tis, either. It's one of the great mortal mysteries, like how does a baby grow, or how long is a man's life. I do know this, however. It's so big, that there are rooms in this building that no man will ever see." He looked at Fred again and smiled his sad smile. "So you see sir," he said, "most of the time, I don't really know what it tis that I'm up to."

Fred smiled. Old Bill is a charming man, even if he is a little eccentric, he thought.

The elevator stopped and the doors slid open. "Take care of yourself, Old Bill," he said.

"It's been a fine honor to have known you, sir." The doors slid shut.

That was a strange thing to say, Fred thought. He wondered if Old Bill was retiring or something. He looked around. He wasn't familiar with this section, and the dust was so heavy he couldn't see very well. He started walking, looking for a wall map. He saw one down an adjacent hallway and walked over to it.

As he stood puzzling over it and wondering if he was on the right floor, beady eyes watched him from a cracked doorway.

Buster turned his head and hissed. "Hey boss!" he whispered hoarsely. "Come look at this!"

Bull walked over and peeked out through the crack. "Well, well, well," he said.

He grimaced with distaste. He disliked violence, but he thought that it was a man's duty before the Lord to kill niggers whenever he could.

"Round 'em up," he said.

In a matter of seconds, the men and horses were assembled in front of the door. Bull pulled out his tobacco pouch and looked into it. Empty. He felt the first twinges of madness creeping into his brain. He scowled.

"Let's go," he said.

Suddenly Fred heard a loud noise of horses galloping and men shouting and guns going off behind him. Before he could turn his head, he was knocked over by two hands, one clamping down on each of his arms. The hands held him up painfully, keeping him from falling. He staggered to his feet and looked around.

His arms were being held by two men on horses. They were big, brawny, and dirty, and they were dressed like old-time cowboys, six-guns, ten-gallon hats, and all. They were grinning.

Fred tried to laugh, nervously. "Say fellas, if this is some kind of a joke ... "

Fred could see that there were several more men on horses behind him. Suddenly one of them rode around in front of him, a massive, stern-looking man who wore what appeared to be a priest's

vestments underneath his sheepskin coat and ten-gallon hat. The man looked him up and down.

"Them's nice clothes for a *nigger*," he said. "Where'd jew git them pretty clothes, nigger-boy?"

Fred began struggling to free himself. "Let me go! Let me go! I swear if this is some kind of damn Halloween prank ... "

Bull reached out and slapped him across the face, hard. Fred reeled with shocked, painful surprise.

"You know what we do with sassy niggers, boy? You about to find out."

Fred struggled to his feet again. He tasted blood flowing into his mouth, felt his lip swelling. Panic started to rise in his throat. "I'll call the police!" he said. "I'll have you thrown in jail!"

Bull motioned with his head. Fred felt himself lifted off his feet and placed on a horse's back, sitting backwards. His hands were twisted behind his back and he felt his wrists being bound with rope. His mind was screaming with fear. He couldn't accept the reality of what was happening. A noose, hanging from one of the bare girders, was looped around his neck. "Help ... don't ... please don't ... " he stuttered.

Bull rode up to him and leaned over until his face was glowering just a few inches from Fred's. Fred could feel the man's hot, foul breath.

"Got any chewin' tobacco?" said Bull.

Fred shook his head. "Don't ... "

Bull slapped the horse's rump.

Jim pulled his mail out of the slot and began to leaf through it. The first thing he came across was a memo. It said:

METAMORPHOSIS

TO: **ALL CORPORATION EMPLOYEES**

Re: **Suicide on Corporation premises**

Earlier today, one of our employees was found dead, hanging by the neck in a part of the building that is under construction. There were large quantities of what appears to be animal feces in the area. Apparently, this person smuggled some kind of large animal, probably a horse, into the building, and used it plus a length of rope fashioned into a noose to kill himself. The animal has not been located as yet.

Obviously, much time and money will need to be spent cleaning up the mess, not to mention training a replacement for the dead man, who did not notify his supervisor that he was planning to kill himself.

We wish to stress that it is in strict violation of Corporate Policy to kill oneself, or anyone else, on Company premises. If you desire to commit suicide, you must obtain written authorization from your supervisor at least ten (10) working days in advance, and you must NOT do it on Corporation premises.

Anyone caught violating this policy will be severely reprimanded.

ORSON J. HARRIMAN AMETHEUDICUS RADCLIFFE STOCKBRIDGE XI
Vice President in Charge of Authority

As Jim read the memo, a sour feeling of guilt pooled in the pit of his stomach. He remembered Bob staring stupefied at a pile of horse manure.

We could have done something, he thought. We could have stopped him. Instead, we went and drank coffee.

He walked over to his cubicle and picked up the receiver of his phone. He called Bob's number.

"This is Bob speaking."

"Bob, this is Jim. I've got to talk to you."

"Jim, I had the strangest dream! I dreamed that you and I were in an elevator, and there was this pile of horse manure … "

"I'll be over in a few minutes." he hung up the phone. Several minutes later, he was walking into Bob's cubicle.

Bob was sitting at his desk. He looked like hell. He had bags under his eyes, his hair was mussed and his suit was wrinkled. A dozen or more empty styrofoam coffee cups littered his desk. He looked sad.

He looked up at Jim. "I don't understand life," he said. "Have you got any coffee?"

Jim tossed the memo on his desk. "Have you seen this?" Bob picked it up and started reading. After a minute, his face lifted into a childish smile, and he looked at Jim. "Horsies!" he said.

✆

Bull was on his horse, galloping full speed through the long musty corridors. He was infuriated. After the thrill of killing the nigger had worn off, he had realized that he had been too hasty. There was valuable information he could have extracted from that boy, such as where they were and how to get out. He was sure they had been here at least twenty-four hours now. twenty-four hours without food. twenty-four hours without chewing tobacco. Bull's grip on reality was slipping, slipping.

He had stopped looking for an exit, a stairway. He had ceased any rational activity, and no coherent thought formed in his twisted mind. He rode wildly through the hallway, viciously digging his spurs into the sides of his horse.

Suddenly he jerked the reins and stopped. He leaped off of the horse. Some impulse made him want to enter one of the rooms. Savagely, he twisted a doorknob, threw the door open and stepped inside.

He stopped. All of his manic thoughts were stunned into silence.

It was big, big enough to enclose three smaller offices. One wall was mostly glass, and it looked out over the west side of the city, with skyscrapers giving way to industrial parks giving way to shopping malls giving way to supermarkets giving way to freeways, ad nauseam, as far as the eye could see. A yellow cloud of smog hung like guilt over the whole scene. Bull stared out over a panorama.

Slowly, Bull's brain began twitching incoherently again. He was stupefied. There was nothing in his little mind to which he could compare what he was looking at. He tried frantically to find something in his memory that he could use as a reference point, but found nothing.

"Lost," he mumbled, to himself. A trickle of drool dribbled out of the corner of his mouth. "Lost ... I'm loooOOOOOST!!" The last word rose into a scream, and Bull picked up a chair and started smashing windows, screaming "MAMAMARY-MOTHEROFGODFUCKINGJESUSICAN'TI CAN'TICAAAAN'TI'MLOSTI'MLOOOOST!!!" After he had smashed the last window he threw the chair out of it as hard as he could. He stuck his head out into the stinging cold air and watched the chair fall, dwindling until he couldn't see it any more. Then he was seized with vertigo, and he recoiled violently, crumpled into a fetal position on the floor, and began throwing up.

✆

Fiona hesitated in the middle of a paragraph. She couldn't read one of the words in the sloppy penciled draft of the letter she was typing for Jim. Her plucked blonde brow wrinkled as she studied the word intently. Finally she gave up and carefully pulled the draft off of the little metal easel.

Fiona was twenty years old. She had a face that belonged on a model and a body that figured prominently in the wet dreams of the middle-aged men she worked for. She was beautiful and bright, with just the right amount of innocence to be startlingly sexy. And she was naively unaware of any of this. Her feathered blonde hair hung halfway down her back, and it swung gently as she walked around her desk, over to Jim's cubicle.

She hesitated in front of Jim's desk. Jim looked miserable. He sat with his head in his hands, staring down at a sheet of paper. His burgundy tie was loose and crooked. It was the first time Fiona had ever seen him when his tie had not been perfectly straight.

Fiona cleared her throat. "Mr. Jim?" she said. Jim slowly looked up, his eyes following the gentle curve of hip, the flat stomach, the breasts straining against the fabric of her blouse, the tanned skin of her neck rising up to meet the flawless face, finally to rest on the brilliant blue eyes.

Fiona felt uncomfortable under his stare, and, she admitted to herself, a little excited. She had been attracted to Jim since the day she had first seen him. He reminded her a little bit of her father.

She cleared her throat again. "Jim," she said, "I can't understand this word here." With that she moved around to Jim's side of the desk. She put the penciled copy down in front of him. Jim calculated that her breasts were about fourteen inches away from his face.

Jim picked up the copy and looked at it for a moment. He couldn't think of anything but the horse manure incident. "Never mind that," he said. "Fiona, I want you to do something for me. You know that section of the building that's under construction? I want you to go to this office," Jim handed her a slip of paper with some numbers written on it, "and see if there's anything ... unusual up there."

Fiona smiled. "It sounds pretty mysterious."

Jim tried unconvincingly to smile. "It's no big deal, really. It's just that I was up there earlier, and I saw some unusual things that I'm checking in to. So if you could do that it would be really helpful."

"Okay," Fiona said with a bounce. "I'll be back in a little while."

Jim watched the graceful swing of her hips as she walked away.

❦

Fiona walked out of the elevator. There were men working in the hallway, cleaning up the floor. They stared at her as she walked past. She smiled and said hello. Then she began the long walk to Fred's office.

It took her fifteen minutes to find it. Along the way she saw piles of horse manure and hoof prints. She thought of the memo she had seen earlier. Maybe I'll get to see the horse, she thought with excitement. She loved horses.

As she turned the corner, she saw it. It was a big, beautiful horse, white with silver gray speckles on its muzzle and mane. It had a beautiful saddle and bridle. Fiona was tempted to jump on it and ride down the hallway, just as she had ridden on the beach at Half Moon Bay when she was a child. She walked over to the horse. It was trying to eat the carpet. When it heard her approach, it looked up and snorted uneasily. She began cooing and making baby noises at it. Presently it calmed down and she stroked the big horse's muzzle.

She could see that it was hungry. "You stay here, and I'll bring you some food, okay?" she said in a voice usually reserved for babies. Then she opened the door to Fred's office and stepped inside.

The cold wind stung her face, and she saw that the windows were broken. The wind blew in her eyes, making them squint and water. She felt something brush against her legs and looked down to see that it was a cowboy hat. It was then that she saw Bull, at the other end of the room, cowering on the floor over a pool of vomit.

She rushed over to him. She touched his shoulder. Bull looked up at her weakly.

"ARE YOU ALL RIGHT?" she shouted, over the sound of the wind. Bull turned his head and looked at her dumbly. As he stared at her, he slowly came halfway to his senses and saw what he was looking at. A vicious sneer spread across his features.

Fiona was suddenly struck with alarm. Before she could move, Bull stood up and grabbed her by the arms, pinning them to her sides. He pushed her backwards onto the desk. She felt her rump painfully hit the edge of the desk, and she was forced into a sitting position with Bull holding her arms.

Bull began ripping her clothes. "You're a pritty little thing," he was saying. Gonna give you some gooood lovin'."

She was terrified. Blood pounded in her chest. She relaxed for a moment, letting Bull think she had fainted. Then she rammed her knee up into his groin as hard as she could.

Bull doubled over, retching. He fell to his knees. Fiona leapt to the floor and kicked him in the forehead with all of her might. Bull rolled over on his side. Fiona ran for the door. As her hand touched the door handle, she heard Bull scream. She turned to look at him. He was back up on his knees. Blood was streaming down his face. He was pointing a gun at her. She was out the door a millisecond before the gun went off. The sound thundered in the cavernous halls for long moments as she ran.

Bob picked up the photograph of his family. He sat, slumped over, staring at it. But he wasn't really looking at the picture. He was deep in his head.

Blood was pounding in his temples. His head felt thick and heavy. He imagined that his blood had thickened from not drinking enough coffee. His body had been constantly laced with caffeine for so long that he no longer noticed its effect, like a speed freak. His shaking hands were cold and sweaty. His stomach felt hollow and swollen. His eyes were bloodshot.

He was thinking about his father. His father had died of cancer of the pancreas when he was six. The doctor had said that it was from drinking too much coffee. His father had been a successful businessman, had become rich by selling little

plastic horses. He had not been around very much, but on those rare occasions when he did come home, he would pick little Bob up and promise him that he would buy him a horse someday, and then Bob could be a cowboy. Bob used to dream about being a cowboy. He had seen "Shane" twenty-three times.

He picked up the memo that was lying on his desk and read it again. His mouth formed the word "horsie" silently. He set the memo down and opened a drawer in his desk. He reached in and pulled out a little plastic horse figurine. He stared at it morosely. "Why can't I be a cowboy?" he said, in the voice of a pouty six-year-old.

Then suddenly Bob's eyes widened, and he sat up in his chair. He was having a vision. The white walls of the Corporation seemed to turn into billowy white clouds. Golden rays of sunlight beamed through the clouds, shining on the formica surface of Bob's desk. Then the clouds parted, and God appeared, looking like nothing so much as Charleton Heston in "The Ten Command-ments." Bob was stricken with awe. He leaned back and took a deep breath.

God looked down on Bob and saw that he was unhappy, and took pity on him. And God said, "Why art thou unhappy?" and Bob said, "Because I want to be a cowboy." And God said, "Verily, thou art unhappy because thou canst not be a boy of cows. But I am thy Lord, and I sayeth unto thee, be happy." Then God parted the clouds, plucked seven cowboys out of the distant past, and plopped them down in the bowels of the corpora-tion.

Bob got out of the chair and prostrated himself on the floor. "O thank you God, thank you thank you thank you thank you thank you."

Bob's supervisor, Norma, who had been walking by, stopped and stared at him. "Are you all right, Bob?" she asked. Bob leaped to his feet. "GET THE

METAMORPHOSIS

HELL OUT OF HERE!!" he screamed. Norma hurriedly walked away.

Bob was grinning like a lunatic, his head bobbing frantically. He knew that the world of cowboys was rough and tough. He tore open the top drawer of his desk and started rummaging through it. He quickly found what he needed. A letter opener. He stared at the silver blade for a moment, then thrust it in his pocket and walked away.

Jim was writing a memo when Fiona stumbled into his cubicle. Her normally perfect hair was in disarray, a few strands falling across her face. She was breathing heavily, a look of shock on her face. Her blouse was torn, leaving one of her breasts half exposed.

Jim stood up, and Fiona fell into his arms, sobbing. Her body was hot against his, and as he wrapped his arms around her, he felt that she was soaked with sweat. "What happened?" he said.

Her voice was choked with emotion. "I ... I went up to that office you told me to go to and there was this man ... and he grabbed me and pushed me onto a desk and tried to ... hurt me ... "

Jim pulled away and held her by the arms. "Fiona, calm down," he said. "What man? What did he look like?"

"He was dressed like a cowboy ... there was a horse up there ... "

"A cowboy?" Jim let her go, picked up his phone and dialed Bob's number.

"Bob's desk, Norma speaking."

"Hi Norma, this is Jim. Is Bob there?"

"He left about five minutes muttering something about horses. Jim, do you think there's something wrong with Bob? I'm worried about him."

"Thanks, Norma." He hung up the phone.

He turned to Fiona. "Fiona, there's something strange going on up there. I've got to go up and find out what it is."

"Don't go!" she exclaimed, wrapping her arms around him and pressing herself against his chest. They looked into each other's eyes. "I ... I need you here."

Suddenly her mouth was against his, open and wet, and their bodies crushed together in the heat of passion. Fiona moaned softly.

Jim reached around her and pulled off his wedding ring.

Bob sighed. He had been wandering around the musty hallways for an hour, and he hadn't seen a trace of his beloved horses or cowboys. He slumped against a wall and sat on the floor.

He silently wondered if he was going crazy. Visions of God delivering cowboys, for Christ's sake. "Maybe I'm drinking too much coffee," he said aloud. "Maybe I should take up aerobic exercise, instead. Maybe I ... " he stopped. Off in the distance, he heard the clopping of horse's hooves.

Immediately, he was on his feet and running toward the sound. Just as he reached the place where two hallways intersected, he saw three cowboys on horseback run through the adjacent hallway. He turned the corner and ran after them, screaming "Stop! STOP!!" at the top of his voice.

The three men stopped and pulled around. Bob dropped to his knees, panting heavily. "Cowboys! Horsies!" he said. Then he fainted.

Swede walked heavily into the freezing room, followed by Buster and Mad Dog. Swede had Bob's inert body slung over one giant shoulder. He dumped Bob on the mahogany desktop.

Bull looked at Bob's sagging form and sneered. He had calmed down somewhat after the run-in with the girl, and the blood had dried into a brown crust that ran down the center of his face. "Well," he drawled, "what have you got here, Swede?" The Swedish giant stared impassively and said nothing. He was mute.

Buster, who was standing behind Bull, near the window, huddling and shivering, whined "Why do we have to do this in here, boss?"

"SHUT UP!" Bull screamed. He looked back at Bob. "He got some pretty clothes just like that nigger boy. We gonna have us some fun with this boy." He chuckled.

Buster's whine grew more high-pitched and insistent. "But it's *cold* in here, boss!"

Bull whirled around. "I said shut UP!!" With the last word he raised up and kicked Buster squarely in the chest, sending him reeling backwards and out the space where the window had been. Buster's piercing scream dwindled in volume and disappeared.

Bull was red-faced and furious. "ANYONE ELSE GOT ANY QUESTIONS?!" he screamed, hysterically. No one did. Angrily, he pulled his little red book out, crossed himself, and said, "He was a sniveling little son of a bitch. Amen."

Bob let out a snore.

Bull turned back to Bob and backhanded him across his peaceful face. Bob yawned and opened his eyes. He raised a hand to his cheek. "Hey," he said groggily, "that hurts."

Bull grabbed Bob's tie, pulled him up off the desk, and stuck the barrel of his gun in Bob's left nostril.

"It's gonna hurt a hell of a lot more less'n you start giving me some answers, quick!"

Bob smiled a silly sleepy smile, all the sillier because he had a gun up his nose. "Cowboys," he said.

Bull shook Bob. "Where are we? What is this place?!"

"This is the Corporation," slurred Bob. "God brought you here in my dream."

Bull smashed Bob across the face with the gun, sending him sprawling across the floor. Then he grabbed Bob again and jammed the gun against his head. "You gonna talk straight, or do I have to blow your fuckin' brains out?!" he yelled.

"I am talking straight," Bob pouted. A trickle of blood ran down his chin.

Bull threw him down on the floor and kicked him in the ribs. As Bob clutched his chest in pain, Bull leaned back on the desk and said, "Mad Dog, beat the shit out of him. I don't wanna kill him 'til we git some information out of him."

Mad Dog grunted and rolled up his sleeves. He walked over to Bob, picked him up by his shirt collar, and punched him in the face. Bob sprawled and slammed against a wall. He remained standing, shakily. Mad Dog laughed and walked toward him.

Suddenly a transformation seemed to come over Bob. He gritted his teeth, pulled something out of his pocket, and assumed a fighting stance. This only made Mad Dog laugh harder. He grabbed Bob by the shirt and prepared to hit him again. As he did this, Bob took the letter opener he was holding and fiercely drove it into Mad Dog's heart. Blood spurted all over his starched white shirt sleeve.

"Shit!" was the last word Mad Dog Mulhauley ever said. He fell to the floor with a thud.

METAMORPHOSIS

Bull and Swede stared in disbelief. Bob pulled a gun out of the dead man's holster and pointed it at Bull.

"Now I can be one of you, right? Now I'm tough, right? Now I'm a cowboy, right? Right?"

Swede made a move to draw his gun. Bull put his hand out to stop him. "I know a good fightin' man when I see one," he said. "Sure you can be one of us."

Bob smiled. It was the happiest moment of his life.

Bull smiled back at him. "Now how do we get out of here, fightin' man?"

✍

Jim stepped out of the janitor's closet, straightening his tie. He decided to go to Bob's office and see if he was back yet. As he walked down the hall, a familiar electric whine droned to a stop behind him.

He turned to see Old Bill leering from the seat of his mail cart. "Want a *ride*?" Bill shrieked.

Jim gulped. "Sure," he said. Jim eased himself onto the orange seat. Old Bill grinned/sneered at him.

"Things getting weirder and weirder around here, sir," said Bill. "One shouldn't be walking these halls alone. Where you be the taking ta-day, sir?"

"Mr. Bob's office," said Jim.

"Ach! As it was, so shall it ever be," Bill exclaimed. "However, you won't find Mr. Bob at his desk right now, sir. He's taken up with some rather surly gentlemen up in the construction area."

Jim frowned. "Surly gentlemen? What are you talking about?" The cart stopped in front of the elevator. "Take me up to the construction area!"

Bill leaned over and punched the UP button. "Do what you will, sir," he said, then, leaning toward

Jim, "but between you and me, I wouldn't strongly advise it."

Jim was deep in thought. "I've got to find out what's going on up there. This just keeps getting weirder and weirder."

"Weirder and weirder," Bill repeated.

As they rolled into the elevator, Jim suddenly realized that he was sitting next to Old Bill and had hardly said a civil word to him. "I'm sorry, Bill," he said. "I've been all wrapped up in my own thoughts. How are you?"

Bill suddenly looked at Jim seriously. He seemed to turn the question over in his mind. "When you asked me that question before, I warned you not to ask it again," he said. "But now I think a straight answer deserves to be given."

He leaned over, right up close to Jim's confused face. "Everything returns to nothing," he said, "light dissolves into darkness. Everything that lives, dies, and everything that is warm turns cold with the coming of night." The tone of his voice became more intense as he spoke. "In time, everything that moves will become still, and return to the center, from which it came. And then, this all that is nothing will wait to be born again."

The elevator doors slid open. They were on the construction floor. Jim was speechless. He stepped off the mail cart and wandered absently into the hall.

Old Bill smiled after him. "Goodbye, Mr. Jim," he said.

✍

Bob had chosen Buster's horse, Buck's hat, and Mad Dog's guns. He was galloping down the hall, with Bull, Swede, Jed, and Little Whipper close behind. He had never felt more alive in his life. He was leading the way to the elevator. He would take them out of this building, out onto the streets, and then they would really raise some hell. Bob chuckled to himself. In his mind, he recited a poem he had learned when he was a little boy:

No more pencils,
No more books,
No more teachers' dirty looks.

He assessed the condition of the men and horses. They were all very hungry, and the horses were in bad shape. Bull kept babbling desperately about chewing tobacco. Maybe we'll rob a grocery store first, he thought. Or maybe a drug store. Or maybe a ...

Suddenly he waved his arm for the men to stop, and woahed his horse. Jim was standing ten yards in front of him.

Jim walked toward Bob, looking puzzled and bewildered. "Bob, is that you? What's going on here?"

Bob pulled a gun out of his holster and waved it at Jim. "Now you just shut the hell up, Jim," he said. "This is my life, and I'll do what I want to."

Bull and Jed had jumped off of their horses and were walking toward Jim. They each grabbed one of his arms and held it. "You want us to take care of him, Bob?" Asked Bull.

"Bob, is this some kind of a joke?" said Jim.

Bob scowled. He didn't know what to do. "Tie him up and we'll take him with us," he shouted.

As Jed tied his hands behind his back, Jim was shouting, "You can't do this, Bob! What will your wife and kids think! Snap out of it, Bob!" Bull slapped him and told him to shut up.

Presently they had bound his wrists, and he was sitting behind Bob. After a few minutes they were in front of the elevator. Bob pressed the DOWN button.

"Now remember," Bob said to the assembled men, There's only room in this elevator for three horses. Bull and I will go down first. As soon as we get in and those doors close, Whipper, you hit this button here. After a little while, these doors will open again, and you men get inside. There's a panel of buttons on the inside of the elevator. Press the button on the bottom left side of the panel. Bull and I will meet you in the garage."

The elevator doors slid open, and Bull got into the elevator, ducking his head as he did so. Bob began to follow him in.

Jim, sitting behind Bob, was thinking frantically. He didn't know exactly what was happening here, but Bob was obviously in one of his moods. What had always worked before?

"Want some coffee, Bob? Coffee? How about a nice cup of coffee?"

Bob stopped. "Coffee? Coffee?"

"Git in, Bob!" yelled Bull. "I'm gonna push the button and start this thing! Let's see ... bottom left side ... there." He pressed the button. The big steel doors began to close. They bumped against the sides of Bob's horse and opened again.

Bob appeared to wake up. He shook his head and moved into the elevator. The doors closed, and the elevator began to move.

Bob turned around and knocked Jim off of the horse. Jim painfully hit the floor of the elevator, knocking the wind out of him. Bob was furious. He drew his gun and pointed it at Jim with a quivering arm. "You god damn bastard! You tried to trick me!!" he bellowed. Jim sat on the floor trying to catch his breath.

Bull smiled. "Don't worry about that little varmint, Bob. Just think of all the fun we're gonna have lootin' and rapin' and killin'."

METAMORPHOSIS

Bob grinned a malicious grin. "Yeah! Lootin' and rapin' and killin'!" The elevator stopped. Bob looked at Bull gratefully. "You know Bull, this is the beginning of a whole new life for me."

As the doors began to open and water began spraying through the crack, the horses started to thrash about wildly. Then billions of tons of sea water gushed down on them, spewing them out of the elevator and onto the sandy ocean floor. The last thing Jim saw, before the water crushed his lungs, was Bob,

      smiling happily,

      riding his bucking bronco to nowhere.

ORIGINAL WATERCOLOR BY
BILL FRIEND
AaaRrrrr! ♂ *SBNoM*

METAMORPHOSIS

VISIONS

METAMORPHOSIS

Mending Things That Have Long Been Broken

David Rubine

Rod Warriors ♂ SCMC

The wind gives a voice to the pine needles.
The heart of this valley surrounds me like an invisible splint mending
things that have long been broken.
The wind that gives voice to this valley also gives voice to this body.
Broken and silent, what is that invisible weight that contains the wind?
I am unaccustomed to a truth that longs to be known.

The wind or truth, it blows through my arms.
What could they carry that would let them know
the strength of other beings?
What weight does this valley carry other than the thoughts of men who
burden it with meaning.
I am alone and the rock doesn't care.
If this is true who really knows what I fear most?

I need people.
Who will keep me from being swallowed up by that which is more
mysterious than I?
Mystery upon mystery; which knows which?
The one that loves or the one that is.
I walk alone through this ancient valley, there are not two of me.
What then are those sounds that echo within the context of my skull
and make sense of the wind?
Who will keep me from being swallowed up by that which is more
mysterious than I?

These questions only grow.
Multiplying and covering this valley with trees, lichen and moss.
Mystery upon mystery, that joy is mine — to see what grows.
That is how I know what to protect and what to destroy.

That joy is mine.
When you see it let the trees stand apart and the wind whip by
so I can sing what is mine and hear songs which aren't.

Original photograph by
Moe Rubenzahl
Phoenix ♂ SBNoM

The One

JIM KNAPP
More Power ♂ *SBNoM*

January of '89. It has been a magic trip already. Only three days into a week and a half trip that will include Wolf Creek, Purgatory, Telluride, and finally, although I don't know it at the time, heli-skiing in the Rubies.

It's Saturday. I find myself sitting on a rock, at the top of a ski run, in a place that has called me before and still calls me now, Taos, New Mexico. Waiting.

An old Indian ventures forth from the lift shack about 40 yards away. As he approaches, I fear he will kick me off my spot, a spot that to me is the natural spot, I think, based on thirty years in the snow, in the mountains.

"How did you find this place?" he asks.

Not being sure of his motives, I respond, cautiously, "It just seemed like the best place to be."

"It is," he assures me. "I climb to here with my family to picnic in the Summer, because it is the best place. You will see it all today."

He is from the first generation, I learn.

Ernie Blake, the man who had founded Taos Ski Valley over forty years before had made a point of employing local Indians whenever he could. Some of the sons and daughters of those originals work here today. This man was one of the first dozen or so that Ernie had hired.

Ernie had been a maverick with a heart. In the 40's, the Forest Service had said it was too steep here for a ski area. Ernie had built his dream anyway. People had said the Indians would never work. Taos is the second largest employer in New Mexico, almost all locals and many Indians. He had made it a point to consult with the Tribal Elders before doing any developement.

He had been a man famous for many things. For hiding a thermos of martinis somewhere on the mountain, for closing the lifts for a half an hour lunch. For runs like Psycho-path.

He had under-reported the snow conditions. Always. If you ski, you understand. He had dressed as a janitor and swept the sun deck so he could hear what his "friends" said of his mountain.

As you drive into Taos the first thing you see is Al's run. The place where the mogul was invented. It is huge. It is steep. Unless you've been warned, it will send chills down your back. Everywhere there are signs urging you, "Don't panic!! What you are looking at is less than 10% of Taos' famous ski valley." Many people have driven into the lot and simply turned around and left.

They had said no one would come this far, that you couldn't entice the right kind of professionals. Taos is consistently one of the top, if not the top, rated ski schools in the country.

Ernie had had a simple apartment on top of the lodge with a window and a chair so that he could look at his mountain. He is quoted as having said, "I pray there is no reincarnation. Nothing could surpass this journey with which I have been blessed."

As I sit on the rock, my mind drifts back a dozen years to a morning with two of my friends. It was Valkyrie's Glade. It was belly deep smoke. We were free falling in absolute silence at over twenty miles an hour. In 35 years of skiing it was **The One**.

I can't believe that I just "happened" to be here.

It gets very quiet as the lifts stop. I see the shadows first. Then I see them. Down the valley from the East come two A4 Skyraiders of the New Mexico Air National Guard. *On it.*

Ernie Blake, a former member of the Air Guard, has died of pneumonia the week before. Today is the wake.

Up the ridge on the right I hear them coming just before they explode into view. Then it's out over the back bowl slow rolling and over the far ridge clearing it by thirty feet. Maybe.

It's quiet again. A few moments to reflect on this man, this mountain, this dream.

Again I see the shadows first, this time up the valley from the West. Turning hard and hugging the ground they're heading right at me at maybe twice tree top level!

It's probably a simple maneuver for experienced pilots. As if they are one plane they pull the noses up to vertical and right on top of me go to after-burners.

As they spiral upward the leader simply opens his dive brakes and the ashes of The Man I Always Wanted To Be find their final home.

Light

ANDREW CREELY

Chi-Pachagon Express ♂ *SCMC*

Morning comes.
The brilliant blue of this day builds,
layer by layer
like the silent laying down
of seabeds —
changing light
   like bands of silt and sand,
luminous clouds
   like delicate spiraled shells.

These moments that we share —
   this laugh, this look, this sigh —
are they not the same?

The spiraled shell drifts down
   from light to dark,
   from what was and bloomed
to what may be: a million or a billion years,
   the seabed now uplifted to the light,
   the spiraled shell embedded in the common stone
   of memory.

The edges smooth,
the hand closed round — this laugh,
this look, this sigh —
picked up and turned and turned by waves,
found tumbled on this evening's beach,
slipped in the pocket
bound for home.

A Full Life Remembered

DAVID DECARLO
Guest ♂ SBNoM
(Posthumous)

I close my eyes and see bamboo
dark cool jungles
golden yellow bamboo stalks across
deep green
the color of tigers
crouching

ℬ

David lived his life fully. He said, "When people sit around and ponder what they would do if they found out they only had a week to live, they think of all the things that they would have done if they'd had more time. I find myself in the position of only having a couple of weeks to live and I wouldn't do anything differently from the way I have been living my life." David died on March 2nd of 1993 at the age of 44. —**Eric B. Lane**

perspective

RON REINARTZ
NFL ♂ SBNoM

Clouded by LIFE
on a day to day
hour by minute existence.

LIFE becomes
mechanical
superficial
devoid of reality
devoid of nature
influenced by empty routine.

Remove
the persuasive cloud.
Let the beauty of life
fill your spirit.
Become
another perspective.

In the Silence of Night

JAMES LAWRENCE
Guest ♂ EBNoM

"Hold still! Quit backing away!" There is no breath left. In the darkness of night, it is becoming blacker still. "Hold still, you little shit!" I can't. Or rather, my body can't. Not if it's to survive. Gagging. Choking to death. Tonight he stops before I black out. And so, I'll remember some of this, decades later.

ℬ

Radio on. Car motor humming beneath. I secretly dread these sounds. We're going downtown. To the hardware store. His eyes glaze over. The car pulls over. And again, my body flails helplessly until it goes limp, and daylight becomes darkness at 5 p.m. in July. The last sound is that radio ....

ℬ

He is taking me somewhere. I sense something different than the other times, and a dark ball of terror looms beneath my resignation. We arrive. There are others. "Don't talk! Do what you are told!" The blackness comes over and, somehow, a part of me lingers on, driven by the same urge to survive as before. And somehow, I survive.

Sacred Moments

DAVE DEESCH
Cool Jerks ♂ *SCMC*

Nothing is more inspirational for me than experiencing a sacred moment, which is observing nature in its simplest forms. These precious few minutes that I have had in my life remain the most special, soul lifting, and humbling encounters. These experiences always put life into perspective for me and certainly help me re-evaluate priorities. Two examples follow.

While backpacking in the Marble Mountains with my team, the **Cool Jerks**, I had one of the most spine-tingling moments of my life. We horse packed eleven men and boys to English Lake for nine days of fishing, hiking, and blatant loitering among nature's greatest gift, the wilderness. On a hot afternoon of the second day I took off, by myself, to a meadow a couple hundred feet above the lake to take a leak. While pissing I noticed a young doe grazing behind some manzanita, amongst a field of lupine and other wildflowers. Luckily, my good friend Tim Corcoran taught me the basics of stalking just a few weeks previously. I was a good fifty yards away and very quietly and slowly began to crawl and slither along the ground towards the deer. After at least thirty minutes I was behind the very same manzanita thicket where the doe grazed. I lay motionless for at least five minutes taking in the sounds of the doe chewing, birds chirping and just a few glimpses of the doe through the branches and leaves. She was no more than eight feet away. I wanted more.

I must have spent the next ten minutes slinking along the ground toward a lower part of the thicket to catch the view I so desperately wanted. Ever so slowly I rose my head. There she was! Less than four feet away. Our eyes met, and I was shocked she wasn't startled. I thought I had died and gone to heaven. My heart was pounding, and I was breathing quite heavily when a twig snapped under my foot. To my surprise a three point buck arose no more than six feet away. I stood motionless for what seemed like several minutes, but in reality was probably seconds. I stared directly at the buck, his nostrils flaring and eyes bugged, his ears pointing up and forward. He looked straight at me and instantly pranced and hopped away and up the mountain with his mate close behind. I laid there another ten minutes silently, eyes staring at the cloudless sky, and took in the moment. What a gift! I hadn't been carrying my camera, but to this day the memory of those moments are quite sufficient.

Another encounter happened just a few weeks past. I've just taken a summer position as a park aide at Loch Lomond reservoir in Ben Lomond. Part of my "work" is boat patrol where I cruise the lake in a boat, mostly checking on people's safety. One of the benefits of the job is being surrounded by all sorts of wildlife if you keep your eyes and ears open. An osprey was hovering above, circling the sky and looking for a snack. He screeched and dove straight down, wings bent back, and hit the water with his talons extended and emerged with a trout dinner. He flew immediately to a snag, carrying his meal parallel to his body, displaying his aerodynamic skills. He screeched and ate. I watched in awe and felt quite blessed.

May you all encounter sacred moments!

HO! *Mitakuye Oyasin!*

Writing And Transformation

DENNIS SIDES
Happy Men ♂ SBNoM

I've dreamed about writing a novel all my adult life, and probably even before that. I kept waiting for the "right" (write?) time, but as the years went by I began to wonder whether that time was ever going to happen. So, on January 1, 1992, I sat down at the keyboard and typed the words "CHAPTER 1." Sometimes you just gotta say, "What the fuck."

The first part of the transformation was, well, fun. From the get go, writing was a blast. I really got into creating characters and throwing situations at them. Novelists must have a "God-complex" — they just go to a lot more trouble than most people to create a world they can rule. I couldn't wait to see what would happen next, even on the days I didn't get home till 11 p.m. I noticed I began to sleep less — I went from 7 hours a night down to 5. Some friends expressed concern about my health, but I was doing fine — I was "on my path." I had often wondered what that phrase meant; now I was living it in my gut. I was a man on a mission.

The next thing I did was truly scary. I made my writing public knowledge. I told everyone — family, friends, team, co-workers, people in line at Safeway — that I was writing a book. (By the way, doing this article was a bit of a stretch, too). I'd like to say I've always been a deeply committed man, but the truth is I've been a "don't push the river" guy, hanging back until I saw how it was going to come out before I ventured in. This time I put my balls on the table. The family jewels were going to smell pretty funky if I bailed out now. I declared my writing because I was finally wise to, and fed up with, my old tricks. This was my way of lashing myself to the wheel and sailing straight into the storm. But then a curious thing happened. I found the declaration actually helped me, and my fears about commitment and exposure just made the process more real. Once I committed to it, I became empowered by the work. I was verifying Rule #1 about commitment: "100% is easy, 99% is a bitch."

By the time this article appears, I will have finished draft number 4 and given out a few copies for evaluation and criticism. No one has yet seen the 384-page book, although I have read a couple of chapters to my team. It has helped me to write with them in mind, connecting my writing to their reading. In no small part I am taking on this challenge as a result of direct and indirect support from these men. I am inspired by the incredible things they are accomplishing in their lives, and some of that has finally rubbed off on me.

I've needed something to anchor me, since the journey into the book has been quite arduous in some ways. In the old days, deep-sea divers could fall into a state of nitrogen narcosis, "rapture of the deep" they called it, and not want to come up. So it is with writing. I've submerged myself in a world that is often as real as the one I'm actually living in. Scary stuff. I'm developing a healthy respect for novelists, if only for their ability to keep from becoming totally schizo.

I grew up in Indiana, and I never forgot the quote from the Gary Post-Tribune sports banner — "On the fields of friendly strife are sown the seeds, which in other years, on other fields, will

METAMORPHOSIS

bear the fruits of victory." It took me a long time to realize they weren't just talking baseball and football. The "friendly strife" of this book has generated energy in the other fields of my life. During the past 16 months I've kept up a 50-hour (and sometimes 60-hour) work week; spent a month or two searching for, moving into, and furnishing a new place to live (I'm on a first-name basis with everybody at K-Mart); test drove about a hundred new cars and actually bought one; ran 15 to 20 miles per week; and, last but not least, fell in love twice. So much for my old stories that I never had time to accomplish anything.

I am not the same man I was before the book. There's a belief in myself that wasn't there before, a visceral sense of my own personal power and discipline, a great feeling of accomplishment that is more connected to the process than to the result. The book is already a success — a big win for me — even if it ends up in publishers' slush piles right along with all the "It was a dark and stormy night" books. You write, or paint, or create music, etc., because you love it — the doing, the striving, the process are its own rewards.

There's one curious footnote to all this. During the year before I began the writing project, I got back into running. At the time I thought it was just because I was badly out of shape and couldn't remember anything anymore. I also decided to give up alcohol for three months, at the suggestion of a team member, and I threw in caffeine for good measure. As of this writing I haven't had a drink in two years and continue to get my kicks from decaf. I'm still running and, incidently, getting a lot of my best ideas in the process. Two years ago I didn't know why I was doing these things — they certainly weren't connected to the idea that I would actually be writing a book. They say hindsight is 20/20 and it looks like my transformation through writing

began then, subconsciously, on the physical level first. Maybe this was my body's way of preparing me for what I had always considered "mission impossible." I'm not suggesting we have to become monks to get what we want out of life. The incidence or coincidence of these events is just more of life's great mystery to me.

I'm looking forward to a break in the action — I think every transformation process needs a "period of adjustment" where you sit back and say, "Wow!" I plan to knock on a few doors and see if people still remember me, maybe even *read* a book or two, and take some time just looking out the window. Then it's back to polishing the prose and seeing if anyone wants to publish a story about a wise-cracking guitarist in a rock band, a modern-day Ahab slinging a Telecaster instead of a harpoon. I was incredibly naive in the beginning. I thought the whole process would take a year — "write it, revise it, take the highest offer" — nothing to it. Now they tell you it's more likely getting killed by a terrorist than getting your first book published.

I don't know whether writing is my destination or just part of the journey to whatever's next. It doesn't really matter. After four years of being on a men's team I am "trusting my gut," and that's enough for me. As the great philosopher Elvis Presley said, "You gotta follow that dream wherever it may lead you." And, at the end of the rainbow (which is merely sunlight going through its own transformation process), I hope to find — the man I always wanted to be.

Round Table

RICHARD LEHRMAN
Guest ♂ EBNoM
(posthumous)

In 1974-76 I lived communally in a wonderful old San Francisco Victorian with several people, among them single mother Pat Lehrman and her 13-year-old son, Richard. I enjoyed Richard's company. He was bright, funny and inquisitive. He and I would pal around together and sometimes go on little adventures. In retrospect, this friendship was perhaps the closest I ever came to being a parent or mentor.

After '76 we all went our separate ways. Richard and I would talk on the phone or get together for coffee once in a while — not often enough. On September 23, 1990, Richard Lehrman died a needless death, leaving a great potential unfulfilled. He was 28 years old. Two weeks later his son was born: Ilya Alexander Rueven Lehrman. I am submitting this poem in Richard's memory, and as a legacy to his son. —**John Timothy**

I

Will or not to will, that is the race
for time; the ill-fated childhood.
Enchanted folklore of sorcery and knighthood
all shared at my Round Table timetable
trance sure-fire. Run through the woods
where one-horned white beasts mingle
Goodness thru white dove feather
to strain the Goodness from the naughtiness,
to hold peace with the naughtiness
that afflicts all our souls. Bless us ...
Given an Eternity isn't much time,
prosper from the pool of knowledge,
curb the Anger and Arrogance and Ego.
The Pen is the Mightiest Sword
toward fighting Evil Ignorants
bending sound around and around.
Given time on Snowy White
caps could spark timbre
in the heart, a rainbow
in Sunlight, close your eyes,
a rainbow in the dark.

II

Instant gratification. Feelings bound up
unwind in a cascade of fluctuous feeling.
Contentment conjures a spark of vitality,
a rainfall of hearts, a sheet of lace.
Cry baby cry, your second coming —
coming of the Queen — God save the Queen
Be in. In is to be. To be is in.
A be-in — to be at the count of three,
Everybody start to be free of guilt.
Locked so tight. Pounded in for so many
years. Free at last after an Eternity.
Sew the scars as you look into the stars.
The space in your eyes, the tears way back
when; the smile of someone down.
Help yourself so you can Help me see you
as a broadsword of Hope, willful Glee,
destined light shining Bright shine tonight.
Carry on that torch of self-worth
in the shadows of the one-horned beast.
Pan plays his flute to three wood nymphs
content in listening to three bars or three notes;
Forest children run and Play Gloriously.

III

Fade to white ... write and write,
forest scrolls, scriptures of fantasy.
Cranes, knee-high in the blue waters
on one leg balanced. Contrasting
Black & White Cranes playing on waves,
Blackbirds forever calling someone,
Deers & Does nibbling on the underbrush,
Pan's flute and Cupid's arrows
play on the feeling — Could you touch me?
Give me a sign thou art divine.
A Round Table has knights black & white;
fair maidens let down their hair to glisten,
to shine amber waves in the Sun,
shore to sandy shore.
Tailspin does a Hawk
whose tailfeathers are dusty
brown orange and red,
pursuing a Possum
with Air Pirate techniques.

IV

Forever is a long time for maidens to wait for their lords.
Shore to sandy shore, tangerine-haired
Ladies long for their Lords of the
Round Table out on Conquest to reign
on foreign soil or fighting in the Crusades
or maybe in the Hundred Years' War, wherever
it is noblest. Fixtured Victims of Honor
Save your Shores from Invading Huns,
forever saving Fair Maidens and Damsels
in Distress from Fire-breathing Dragons,
Potion-wielding Witches, and Scepters held
High by Powerful Sorcerers. All raised
doubt about reality quickly composed
as steel against steel wakens the mind
set on evening the score of the Great
Renaissance. Hark! Hath all been lost
in love better than never to have loved at all?
Lancelot and Guinevere, Robin Hood and Maid Marian ...
nothing little about these loves;
Clean and Pure like Snow White,
Delicate and Frail like Doves.

ORIGINAL DRAWING BY
BILL McCONNELL
Phoenix ♂ *SBNoM*

The Fable Of My Life
or
The Development Of *My Energy*

BOB GRANT
Purple Peckers ♂ SFT

ORIGINAL WATERCOLOR BY
PETER ASHLOCK
Wailers ♂ EBNoM

Once upon a long, long time ago, even before the Great Depression — in a far, far away kingdom called Indiana by its benighted inhabitants — a sperm and an egg met, united and started *My Energy* off on its journey. To his knowledge this happened at least six other times in this family since he had two sisters and four brothers. After nine months of idyllic development in a warm, moist, no gravity environment, he was suddenly erupted into this cold, dry, heavy world.

The minute *My Energy* emerged from the womb he said, "I DON'T LIKE IT HERE. I WANT TO GO BACK!" But there was no escape. Right away he found out he wasn't really welcome in this world. After many demonstrations of this, he developed into *Defiant Angry Boy*. But *Defiant Angry Boy* soon discovered that in order to get the milk and cookies he needed to survive he had to put on a mask. This mask was *Good Achieving Smiling Boy*. This all happened before he could talk.

Now *Defiant Angry Boy* soon discovered it took a lot of energy to maintain *Good Achieving Smiling Boy*.

However, he persisted in trying for the milk and cookies even though it wore him out, and the milk and cookies eventually didn't taste very good. He got into such a habit of this that he kept it up for most of his life. *My Energy*, expressed through these masks, gathered many achievements in the course of time. No matter how hard he tried, though, other people always could feel *Defiant Angry Boy* underneath *Good Achieving Smiling Boy*. This soured all the milk and cookies he received.

In the course of time and many achievements later, *Defiant Angry Boy-Good Achieving Smiling Boy* met a Woman. This Woman, who had started life as a *Little Anxious Girl*, had developed the mask of *Caring Empathetic Girl*. But no matter how hard she tried, *Little Anxious Girl* showed through her disguise. She also found out that it was very wearing to maintain the disguise but didn't know anything different.

Defiant Angry Boy-Good Achieving Smiling Boy and *Little Anxious Girl-Caring Empathetic Girl* fell in love and married. They now have two children and three grandchildren. *Defiant Angry Boy* and *Little Anxious Girl* find it impossible to get along. *Good Achieving Smiling Boy* and *Caring Empathetic Girl* have some success in getting along together, but find it difficult because the masks keep slipping. They are both in the process of developing into *Strong Flexible Adult Lovers* able to convert the energy invested in *Defiant Angry Boy* and *Little Anxious Girl*, and *Good Achieving Smiling Boy* and *Caring Empathetic Girl*, into loving each other and the rest of the world.

METAMORPHOSIS

The Monster Within

JIM WILLIAMS
Guest ♂ SBNoM

When I least expect it, he comes out for a simple day's terror.
Of course he's unwanted, but cannot be killed.
He's my past, present and future.

I hide him from friends, hope they won't notice.
If they knew I had this pet, they would not visit.
But once in a while he comes out unwelcome.
And when he does, I'm left to pick up the pieces.

He's dark, strong, and menacing.
He has a power over me that I wish he didn't.
When I'm not expecting him, I act in ways that befuddle.
Oh how I wish I could tame this animal.

He was born in a home of fear, distrust, and shame.
He flourished on alcoholism and anger.
He assaults with distrust, resentment, and over-bearing control.

For many years I didn't know he existed.
He ruined a marriage, a son's infant years, and a relationship with a family.
Life has improved with knowledge of his existence.
But the battle with him has been exhausting.

When he sleeps all is well.
The world is happy and at peace.
But if I wake him, he will wreak havoc with self-doubt and diminished worth.
He destroys everything important: love, friendship, and self respect.

But I do have a plan.
I've decided to make peace with my monster.
Rather than cage and hide him, I've decided to accept that he's a part of me.
The challenge is to befriend him and let him know it's OK.
For if I don't love him, I can't love myself, and all is lost.

He will always be there ready to attack, but I will be there ready to calm him down.
Over time he will be more manageable.
But once in a while he will get out.
Only love will cause his return.

There is hope for a better life without the monster.
For now he is my pet.
I'll call him Fred.

Dreams Do Come True

MARK GOLDSBOROUGH
More Power! ♂ *SBNoM*

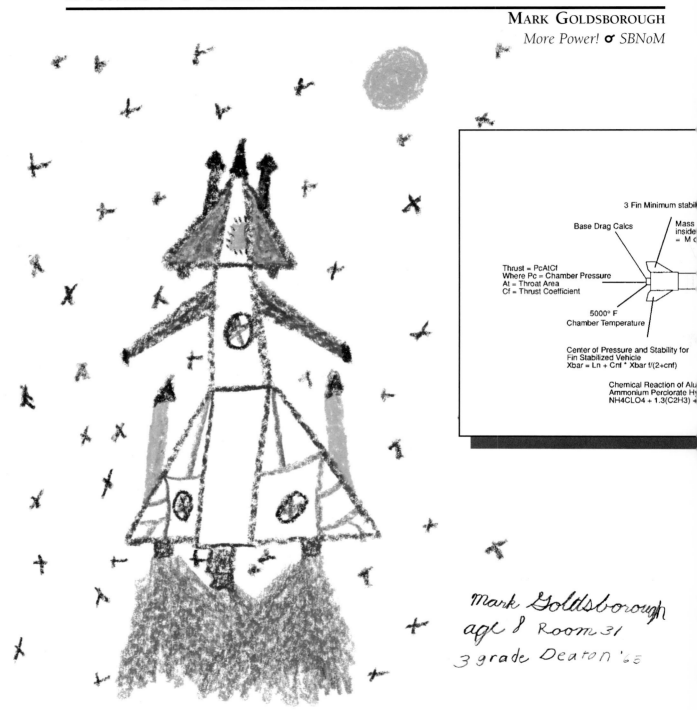

3 Fin Minimum stabil

Base Drag Calcs

Mass
inside
= M

Thrust = PcAtCf
Where Pc = Chamber Pressure
At = Throat Area
Cf = Thrust Coefficient

5000° F
Chamber Temperature

Center of Pressure and Stability for
Fin Stabilized Vehicle
Xbar = Ln + Cnf * Xbar f/(2+cnf)

Chemical Reaction of Alu
Ammonium Perclorate Hy
NH4CLO4 + 1.3(C2H3) +

Mark Goldsborough
age 8 Room 31
3 grade Deaton '65

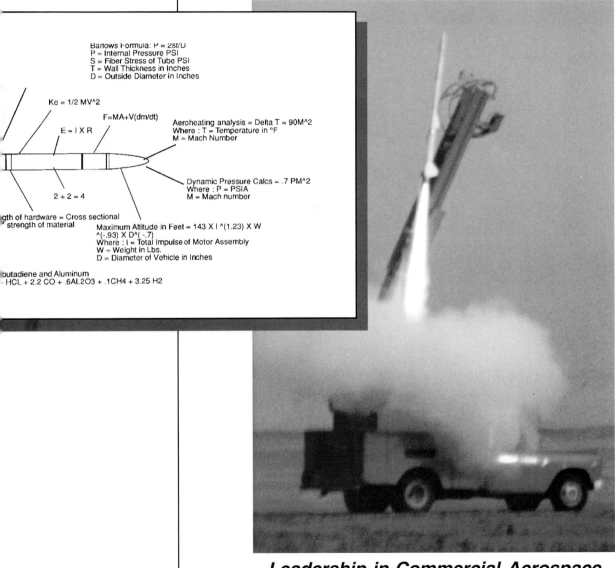

SST *Starflight Space Technologies*

Barlows Formula: P = 2st/D
P = Internal Pressure PSI
S = Fiber Stress of Tube PSI
T = Wall Thickness in Inches
D = Outside Diameter in Inches

$Ke = 1/2 MV^2$

$F = MA + V(dm/dt)$

$E = I \times R$

Aeroheating analysis = Delta T = 90M^2
Where : T = Temperature in °F
M = Mach Number

$2 + 2 = 4$

Dynamic Pressure Calcs = .7 PM^2
Where : P = PSIA
M = Mach number

gth of hardware = Cross sectional
strength of material

Maximum Altitude in Feet = 143 X I ^(1.23) X W
^(-.93) X D^(-.7)
Where : I = Total Impulse of Motor Assembly
W = Weight in Lbs.
D = Diameter of Vehicle in Inches

butadiene and Aluminum
HCL + 2.2 CO + .6AL2O3 + .1CH4 + 3.25 H2

Leadership in Commercial Aerospace

METAMORPHOSIS

A Personal Essay On Change

DAVID WILSON
Purple Peckers ♂ SFT

The Sterling Men's weekend at Camp Swig in the Santa Cruz Mountains in March, 1990, caused a very subtle but noticeable transformation for me. Although I cannot be certain that I would not have changed, regardless of my weekend experience. The changes I experienced perceptively were my relationships to my father and other men that formerly intimidated me. Before I discuss these personal issues I will define a few terms relevant to this essay.

Major transformations are extremely rare for people; small changes in course are much more common. These small changes can combine and reinforce each other to eventually become opposite to a person's original path. In Frank Herbert's book, *Dune*, the author describes time as a collection of infinite paths which all converge at the present. His main character uses a powerful drug with which he is trained to look at the pathways and actually follow one or more mentally to see the future. We mere mortals can neither see the future nor examine our present position based on our past.

The future is not determined in any way for any person just as the present has no conclusive link to our past. It is impossible to retrace our life journey more than one or two decisions back in time because at each decision in the past there have been the same infinite number of possible futures. It is therefore nonsense to discuss our lives in terms of "if only."

There are certain forces in nature which guide all living and non-living things though time. A mountain tends to crumble and deteriorate to a lower level just as water runs downhill. Ants tend to survive and aid in the survival of the colony.

Men and women tend to mating rituals and continuation of the species. At some level all nature interacts: e.g., ants will tend to build a colony out of the direct path of running water or falling rocks, and men and women will attempt to keep ants out of the food supplies.

These "natural forces" are the only predetermined factors in the path each one of us chooses and these can be subtle forces or even forces which can be denied or ignored. The net result and point of this argument is that we are in control, for all practical circumstances, of our past, present, and future(s) as much as any other deciding factor. Our power is very little, only slightly greater than natural forces, but it is enough power to deny us an excuse for whoever or whatever we are.

I chose to take the Sterling weekend because I was looking for someone to tell me what I already knew. Justin Sterling sounded like the kind of person who would satisfy my need for confirmation from everything presented to me at my first open house. I also made the choice to accept at face value just about everything Sterling said about men and women and relationships. For some reason even though it was not new information the way the information was presented felt profound to me.

I came away from the weekend loving and accepting my father in a new non-confronting way. I knew that he had done everything for me with love and respect, and my life had been right by him. If I was perfect then he and everything he had done was perfect. I did not have to explain this new insight to my father although I made a moderate attempt at any rate. He was immediately relieved and assured of where he stood with me. We have had a full and rewarding relationship ever since.

My relationships with other men changed significantly as well. I feel connected to other men since the Sterling weekend. The high anxiety levels I used to suffer around men whom I didn't know well have almost evaporated. It was such a relief to accept that we as men are all the same. It doesn't matter how we appear, what we own or do for work. We are all simple creatures with fragile egos and one essential drive (needless to say what the drive is). Now when I meet a new man I can sniff his butt (figuratively speaking), and do the traditional dance that men do.

These insights I received at the weekend are so deceptively self-evident that they had escaped my conscious mind for the first half of my life. I do not have a perfect life or perfect relationships with men or women but I am encouraged that the direction I have steered is positive. Considering where I have been I am thankful every day (almost) for the opportunities I am constantly receiving to be happy. I may or may not take these opportunities but at least I have the choice.

Camel

JOHN RENISH
AaaRrrrr! ♂ *SBNoM*

Grey buff brown green flat
then a puff of white
like an impossible cloud
looms off the line
where some faint blue
squeezes from the dust.
Coming closer, it's a ship
all plain sail
and liquid brown eyes.
Majestically, she paces
across the desert.

Then and Now

ERIC B. LANE
Quiet Riot ♂ *SBNoM*

THEN (1968)

Twinkle, twinkle star light sky,
I so often wonder why,
Man is just what now can breathe,
Or is it more one can conceive?
Is a man real blood and soul,
Or is he just hard rock and coal?
And in a dream he just shall be,
Tell me, is he really he?

NOW (1993)

Starting on the trail so late in the day,
I don't want to run out of time.
Around the turn, I see the mountains,
So much to see, so much to do,

Another turn, I stop to look,
I feel my throat, I hold my gut,
It is big, so much bigger than me,
So much beauty , so much to do.

I walk the trail, and finally notice,
Spots of color, smell of the air,
Sound of the wind, tingle of my skin,
Time melts away, and I am there.

Summit

David Rubine
Rod Warriors ♂ *SCMC*

Original drawing by
BILL McCONNELL
Phoenix ♂ *SBNoM*

Slow, deliberate.
Inching upward.
Fingernail edge — pull from the center.
Taking in breath — release.
Balancing on top of feet,
to reach — again and again — like lovers surrendered to the inevitable.

Sunlit slabs sweeping upward into the sky,
blue bordered granite horizon.
Undulating waves of polish.
Succession of delicate moves-balance, toe-edge, finger tip biting disconformity.

I can smell the body, the warrior-smell.
Smeared sweat on brow burning the eyes, wetting the hands.
Breathing slowly — opening the eyes to see the faintest undulation or edge.
The rising pulse, anticipation.
Testing the hold, moving, up ... slowly.
Delicately placing the foot on a minute edge — balancing.
Reaching with right hand — nothing ... nothing — a faint flake of rock.
Pulling up slowly, the top — there — one more move.
A thought — I can't — a thought.
Moving — the quiet settles in, breath drops down lower.

Knowing: summit before summit.

Pulling over the top — release, a flood, escape from reasons.
Final surrender to paradise.

The wind carries songs of angels.

My partner coming up, pulling up the slack, I can hear his faint breathe below.
He joins me, says nothing.
We laugh.
The kingdom is at hand.
The voice of angels celebrate our union.
The precious mountains rise up welcoming our return.

We Are The People

STEVEN CERVINE
Rod Warriors ♂ SCMC

In September of 1991, I went on a Vision Quest on Tim Corcoran's land near Mt. Shasta. I was looking for deeper clarity and direction of my spiritual life path. The mountain is a wonderful inspiration, and Tim's guidance and support is invaluable. Spending four days and nights in the woods with no food or other distractions is a great way to get in touch with yourself. Among many other realizations about myself came the inspiration for the words to this song and the desire to begin sharing my music with others.

This song is about what kind of people we are, how we choose to live, and our relationship with the land we live on. It comes from a feeling of intense frustration about how blind I feel our society has been, from the moment we chose to conquer this land and the indigenous people living here, to the inherent wisdom of the land and her people. We must learn to walk in balance on Mother Earth. And as a society we must learn to stop killing off our visionary leaders. There can be so much hope in the blending and harmonizing of dualities, in the integration of modern technologies and the wisdom of the Earth.

And we are the people who will choose the way, and walk the path, and sing the songs for those to come! *Ho Mitakuye Oyasin!* All are my relatives!

WE ARE THE PEOPLE

We are the people who conquered this land
We are the people who take what we can
We take on any challenge to defend
our way of life
Even if our way of life is the
cause of our strife

Land of Truth and Liberty is what
we fought for
But what of the Native People whose heart
and soul we've torn
People with a wisdom to love
simply and true
To walk in Balance on Mother Earth under
skies so blue

Will we be the people to let wisdom
choose our course
Or end up like the Romans, crushed by
our own force
Let us seek the wisdom to make
this country strong
Let's not wait 'til it's too late to admit
when we're wrong

I say let the people live
Yes let the people live!
Hanta yo, wakan tanka (prepare the way, Great Spirit)
Let the people live

Let us return now to the truth from which we built
There's no need to get caught up in a lot of
hate and guilt
"By the people and for the people" is
what we all hear
But you can't kill off the people to
make the way clear

And what of our Mother Earth, to whom
we owe so much?
How long can we go on destroying what we touch?
Mother Earth and her children are our relatives
Let us learn to love our family and tell it like it is

We are the people who decide for today
We are the people who choose to live this way
We must be strong enough to stand up for our rights
We must love each other enough to claim a new
vision in our sights!

An-Nafud

JOHN RENISH
AaaRrrrr! ♂ SBNoM

Bricks ground down to ruddy sand,
Heaped in piles for a giant child's play,
Matched by thousands strewn over land
Where new sun hints of burning day.

A greening furze, both pale and spare
Graces dingles and whilom slopes,
Calling to mind some melon-y hair
Or a balding Martian's age-dashed hopes.

Al-shamal blows hard from Ararat
And scours and shapes all that has grown
From grassy hills where plane-trees sat
Creating new while killing known.

A darker sand of greater weight
Rolls into valleys, a blood-red dust;
Each naked hill seems like the pate
Of a sun-struck djinn transfixed by lust.

When falls the night that steals the heat
And day progresses into dark,
Magenta shades pool sharp and neat
And the moon springs up, skull-like, stark.

The Strength of Desire

BILL HELLER
More Power! ♂ SBNoM

A Holy Man walks through the World
With an erection.
Such an erection,
his pants can't hold it in.

And there is nothing to hide.
For all the World who knows this man
knows his erection is but a hearty symbol,
The symbol of his lust for all of life.

The erection
is a symbol of his energy and desire
to produce in the world,
to make a difference.

In a healed, wholesome World,
Women honor this man,
knowing he completes their pairing.

And in this natural expression of life
and love for life,
all shame and guilt is recognized
for the evil dis-empowering role it plays.

The Men of this World rejoice in his presence,
for they know they are that man, awakened.

A Vision Quest

RONN LETTERMAN
Sperm Whales ♂ *SCMC*

In the wilderness of Northern California, at Mount Shasta, I did my first Vision Quest. For several years I had heard of men from Santa Cruz going north to do Vision Quests. At first these Quests sounded like fun outdoor camping trips. Images came to my mind of campfires, sitting in a small circle for a day or so, sitting Zen style, that kind of stuff. Afterward I could go tell my friends what I did. I would be cool and my status as a seeker of truth would be known.

Other men I know told me about Quests, but they usually didn't say much. I've been part of the Santa Cruz Men's Community for many years. Several of the men invited me to go with them and do the next Quest that Tim Corcoran was facilitating.

What is a Vision Quest? As explained by Tim, the actual Quest time is a process which lasts two, four, or more days. There is preparation time, Quest time, then action time.

For years I've heard and talked about doing a Vision Quest. My opinion is that Vision Quests are a matter of trust, commitment, and honor. I trusted my guides and my innermost self. I committed to living at least four days of my life in a new way, in a new context. I honored the process.

There is a lot written and said about Vision Quests. In truth they are all different. Not just who leads them but how you do them. The time, place, and focus of each Quest differs. The insights a man or woman has will vary.

PREPARATION

Four men I knew said they were going to do the next Quest, along with three other men. That would be seven of us to share our insights and intuitions. But, you know how things go, you plan an exciting adventure then money, work, family, or emergencies usually are the reasons or excuses you cancel.

Two of the guys I was looking forward to being with canceled the morning we were to leave, for familiar reasons. I later understood that when it's your time, it's your time. You can only go on the Quest you go on. There is no right or wrong time to go. You go, when it's your time.

It wasn't until two days before the quest I really committed. Somewhere in the back of my mind a

ORIGINAL PHOTOGRAPHS BY THE AUTHOR

voice was telling me it was time to nurture my spirit. Something was stirring deep in my emotions and I had the feeling it was time to look, see, discover and grow up. No more excuses.

It had taken me years, months, weeks, and, lastly, days to finally make the commitment to do that Vision Quest. During the drive up north I was engaged in asking myself questions and looking at what visions I might experience during the Quest. Also, how my life might differ after I returned home.

Tim told me what items and attitude to bring, to prepare myself for the weather. We were to meet at the base camp on his land, looking east, directly in the morning shadow of snow capped Mt. Shasta. We would spend a day talking, reading, and praying (in our own way) before we walked the land and chose our "Quest Spot."

I won't go into detail about all this, because each Vision Quest leader has their own style of Quest. It was an exciting and wonderful process. We had just to go out scouting and find our own individual spots. We all knew where the base camp was, but as we chose our spots we didn't want to see anyone else's spot. I looked at sunny, shady, view, and stream bank spots and finally chose one that felt really just right to me. I knew where I would be for the next four days. Or so I thought.

During the quest we were to make no fires, eat no food, and drink only water. This sounded "easy" enough. Tim also recommended that we not take things with us we could use as distractions, like radios, walkmans, homework or business books. He did recommend that we take paper and pencil.

We would return to base camp after four days alone. Tim and Mike would look for a "sign" from each of us, every day, in the morning, to make sure we were OK, that we didn't starve or get eaten by bears, hadn't injured ourselves, or run away.

At forty years old I had never been alone for more than two days. And usually there was some form of distraction nearby. There was always something to do, or someone to talk to. I can't remember a day in my life when I've not eaten solid food or drank juice, except when I was sick with the flu. I've mostly eaten three meals worth of food a day or made up for it the next day.

We met at a base camp on Saturday. We had two days to prepare ourselves before we ventured into the forest.

Late that afternoon a woman named Rebecca was returning to base camp after a two day solitary quest. We met her near the campfire. She looked peaceful and clear. Occasionally she spoke and told of her fears of being alone, of snakes, mountain lions, and bears. Later, before dark, we prayed and sang in the Sweat Lodge. We wondered about

METAMORPHOSIS

other people on their own "Quest" in solitude, prison, lost and alone. In the lodge she shared her true experience, her visions, her love. Had you been there you would have heard her. HO! *Mitakuye Oyasin!*

By next morning there were five men ready for the Quest. Each of us were prepared to spend four days and nights alone, by ourselves, fasting in the wilderness. The two other men, who were our guides, were to stay at the base camp.

Some of the animals I saw were ants, bees, bugs, bumble and honey bees, birds, blue jays, butterfly, deer, flies (lots of flies), grasshoppers, grouse, ground squirrels, hawks, lizards, quail, rabbits, snakes, scorpions, spiders, woodpeckers, and of course my self. I ultimately saw creatures, of many sizes and types to this area, dance to their own lives.

The following poems were composed during the Quest. They are the way I would like to express my Vision Quest.

DAY 1 THOUGHTS

MORNING

I look up at Mount Shasta,
the high altitude Bombers
leave their trails,
across the sky.

I realized, there are a variety of flies.
I shall draw one,
under the eye,
of a magnifying glass, Soon.

Have you seen birds fly free,
into the polluted air?

I've begun this Quest.

AFTERNOON

The red tailed hawk flies overhead,
Sun light glistening,
through its patterned feathers.

In the distance, dormant,
a snow capped volcano.

EVENING

I felt wounded by my Father,
abandoned, never knowing it,
Denial, it didn't matter then,
does it matter, now?

I deny the deaths and near deaths
experienced in my life, forgotten.
Sometimes I remember.

Forgotten myths, dreams, memories
I shall never know,
I'm in my space dome tent,
cruising around planetary space.

DAY 2 FEELINGS

Is this not heaven,
the joy, to be,
to appreciate the abundance,
of life.

Blood of pine, dew drops.
Lay strewn, burnt trunks
of an ancient forest,
life abounds,
from their futile ashes.

METAMORPHOSIS

Snow capped peaks,
clouds drift across the Mountain
their shadows cast,
across meadowed plains.

The wind is barely moving
but I hear its breath,
whispered by the trees.

The birds play and sing in the meadow.
Waiting, to see the family of deer,
we all enjoy life.

Sing. Sing, sweet birds,
the shadow of death
is the companion of all living things,
our life but trails for new life.

As sunlit clouds leave shadows,
we return our breath of life,
to the Source.

Mother, father, earth, sky,
cosmic brilliance, coldest space, hottest sun,
life created life.
Do you wonder,
who created who, my god,
who created you, my god?

Day 3 Chow Time

I'm amazed, I don't see
the little critters as food.
Maybe I'm just not hungry
enough yet.

I could eat a dog,
perhaps a human too,
unless he be my son;
then I'd have to be
a Chronos Father.
What would Buddha
say to that?

The earth devours,
all living things,
and waters and births,
all living things.

Sun provides heat and energy
on the surface,
and molten heartbeats resonate;
Where they meet,
is life's-plain of diversity.

I don't think about hunger
as I thought I would.
No Jack-in-the-Box or McDonalds,
or family restaurants
to cue me to mealtime.
I think I'll invite my family
and friends home, for food.

The snake lives in the tree too.
It burrows beneath the earth,
and is found in the crown
of the trees of wisdom and knowledge.
Although it cannot fly,
its cousins do.
We are all of the family of life.

There is evidence of the big cat
and the bear.
Am I prey to become fodder,
or, am I the appetite to devour?

Here Kitty, Kitty and Winny the Pooh,
I have a hungry tummy,
just a waitin' for you.

Here critter, critter don't be afraid,
I'm having you for dinner,
with some pink lemonade.

METAMORPHOSIS

Roast 'em, toast 'em,
eat 'em raw,
but don't let them eat
daddy Ronn's paw.

Stomp it, chomp it, tear it down,
let's eat the earth to the ground.
What does god eat?

Day 4 Bliss

Butterflies dance,
on currents,
of the invisible.
Is that anyway
for caterpillars to live?

I desire a lady,
to share my castle of life;
A knight,
to be a King,
to be forgiven and blessed
by a Queen of Heart,
and perhaps magically touched,
by the enchanted gods.

I sat on a log
a long time today
my butt doesn't hurt
but my back does a little.

It's getting so dark,
I can barely write.
I could use a flashlight
or build a fire,
But that would be artificial,
wouldn't it?

Action

The action that I have taken since this quest has been to co-create *Metamorphosis*, this publication; "Welcoming the Wildman," a safe place to get dangerous with men; and a number of projects, events, and experiences which manifest the "Spirit of Man." I have and will continue to live my life in quest of "Transformation in Action."

Namaste.

Original painting by
Daniel Stolpe
April Fools ♂ SCMC

Thoughts on Transformation

PRESTON NI

Golden Gate Jerks ♂ *SFT*

Transform: To Change. At age twenty-eight, I am in the midst of profound and significant change. When I took the Men's Weekend almost two years ago, it was literally a transformation for me from being a boy to being a man. Over the past two years, as I graduated from college and began working in the "real world," I started to appreciate more and more of what being a man means to me. To me being a man means living a responsible life, holding myself accountable for my own actions, contributing to society, creating and maintaining healthy and intimate relationships, being true to myself, having the courage to change, maintaining a sense of humor and aliveness, and most importantly, following and trusting my heart and my bliss.

On my twenty-fifth birthday I wrote in my journal that I have the next twenty-five years to make something of my life. Today I can honestly say that I have the privilege of living a life which I envisioned not too long ago, and that each day for me is a blessing. I am growing and becoming more aware of my purpose here, of what I have to contribute, and what I must do to progressively realize a future worth living.

I Believe

I believe that love is mightier than hate;
Courage triumphs over fear

I believe that we are only limited by the scope of
our imagination;
Our lives go as far as our vision

I believe that love is to give;
Through giving we reveal the breadth of our
character and the depth of our soul

I believe in magic;
Create new possibilities in the midst of seeming
impossibilities

Finally, I believe that you and I are ONE;
Differences, boundaries, and separations are
merely myths over our common bond

METAMORPHOSIS

Medicine Man

RAY GEORGE
Men of Honor ♂ SCMC

Watch the eagle hunt from the sky
Getting ready to kill or to die
To survive, he does what he must
Perched between the clouds and the dust

See the wolverine in his lair
Not afraid of lion or bear
Knows that he can make his own way
Thrust within his passionate play

Oh, the rattle shakes, the drum begins,
Feel the feathers fan...
Listen to the Medicine Man

Let your arms encompass the earth
Taking every breath, every birth
All you see is not all there is
Far beyond lies the kingdom of His

Initiate your own vision quest
Let the shaman do all the rest
Gates will open at your behest
Will you take the ultimate test

Oh, the rattle shakes, the drum begins,
Feel the feathers fan...
Listen to the Medicine Man

Prove yourself a man among men
Face the truth again and again
You'll know when your challenge is met
Merging with your own silhouette

Wanken-Tanka sheds his disguise
Gaze into your Grandfather's eyes
Learn to see and then recognize
There's no room for your alibis

Oh, the rattle shakes, the drum begins,
Feel the feathers fan...
Listen to the Medicine Man

The Search

DAVID BLAIN
Modern Warriors ♂ EBNoM

It all became so clear
Garden of the heterodox
Pit of Achilles' heel
A masterpiece of vital principle

Violence of the whirlwind
Craving the camouflage
Who's to say the will of Allah
The passing away of embryology

Sour, bitter and balance
A jar of honey
It's all so very simple
But of course, finding TAO

Humans

MATT SORGENFREI
Quiet Riot ♂ SBNoM

Human may simply mean
being different
Like fingerprint whorls

As individuals
we have fears, needs, and hopes
which basically are the same
from person to person

But what makes us human
is the way we touch each other
and reach for the fire inside.

Swirl together our passions
and our dreams
this fire explodes

And from the center of the burning
our children come
to show us again
the glory of being alive.

Warriors

RAY GEORGE
Men of Honor ♂ SCMC

Baby, can't you see what I'm saying
Maybe, you can hear what I'm praying
Don't be among the ones who delay in
The lessons of the lord
I don't like war, I don't like, but I like, warriors
I don't like war, I don't like, but I like, warriors

There's a battle a ragin', outside these halls
No more cast for the curtain calls
There's a mob, and they're waitin', to climb these walls
No more acts till the curtain falls

There's a drama unfolding, state of the art
Every man has to play his part
The rehearsal is over, so we must start
To attune to each other's heart

When will the seed be sown
When will the tree be grown
When will the truth be known
When all people raise their head
And injustice is dead

Now the full moon arises upon the night
Shedding darkness before the light
In the eyes of a Warrior, so clear and bright
An awakening soul takes flight

Baby, can't you see what I'm saying
Maybe, you can hear what I'm praying
Don't be among the ones who delay in
The lessons of the lord
I don't like war, I don't like, but I like, warriors
I don't like war, I don't like, but I like, warriors
I don't like war, I don't like, but I like, warriors

ORIGINAL DRAWING BY
MARK LYELL
Modern Warriors ♂ EBNoM

Warrior

JOHN BUTLER III
AaaRrrrr! ♂ SBNoM

A warrior who thinks he has surrendered
has merely engaged in a strategic retreat.

A warrior who thinks he is giving
has merely called a truce to buy time.

A warrior who lays down his weapons
and his armor is a man.

A man who ceases to compete,
and instead follows his heart,
takes a stand that will create a sage.

Warrior

MARC A. HALPERN
Tatonka ♂ SBNoM

A Warrior is a man who seeks truth in his life
and defends his truth.

A warrior's danger is in being comforted by false truths
and defending them

To know truth from false truth
A warrior must look deep inside
Beyond the illusion
Beyond the ego

To be a warrior
A man must differentiate truth from false truth

To stay a warrior,
He must defend it.

Gillette Man

Rod Warriors ♂ SCMC

Between a rock and a hard place the Gillette Man;
Perfect job, perfect woman, perfect children,
perfect car,
and my own mythical dreams.

Gillette Man and being lazy
hold my balls for comfort, seems so wasted,
only attachment to what I feel;
Lost in space, Penny my first true love,
Oh, I don't think I'll ever recover,
don't touch me there, devil, brotherly love.
Love from my men's team.

✦

Seven trees in a circle
rooted separately.
Leaning toward the center,
collecting the light.

✦

The closeness of love
is painful is fun
is challenging is scary.

The attainment of love
is confusing is joyless
is hard work is scary.

It is not a question
of finding love from without.
It is a question of knowing
the love within.

✦

Who knows these men?
What more is there to talk about
than the silence that beckons.

We stand in open space
our words ripple
with the longing for our real names.

Regularly we meet looking
for something that swims upstream
further than we've ever gone.

✦

Sweaty, hoarse, tired,
drumming heart, drumming spirit,
men together, so much alike.

Building trust, facing fear,
joking and jostling.

Showing up for the dirty work,
drinking around the campfire,
Drumming and doing.

✦

Skip, slide slowly
role jump, grow.

✦

Water, fire, earth, and air
becoming one in an old man's stare.

Brothers divergent yet held as one,
Universe unfolding in the light of the sun.

METAMORPHOSIS

Vision of Men

KIM WELCH
Chi-Pachagon Express ♂ SCMC

Men Stand Strong. They Flex and Beat the Chest.
Men Stand Alone, and look from side to side for Spirit.
Men Stand together and wonder what to do
Men Stand and Howl at the Moon. THEY know what to do!

ORIGINAL PHOTOGRAPH BY
MOE RUBENZAHL
Phoenix ♂ SBNoM

INDEX of CONTRIBUTORS

METAMORPHOSIS

METAMORPHOSIS

METAMORPHOSIS

INDEX OF CONTRIBUTORS